Plant Structure and Development

A Pictorial and Physiological Approach

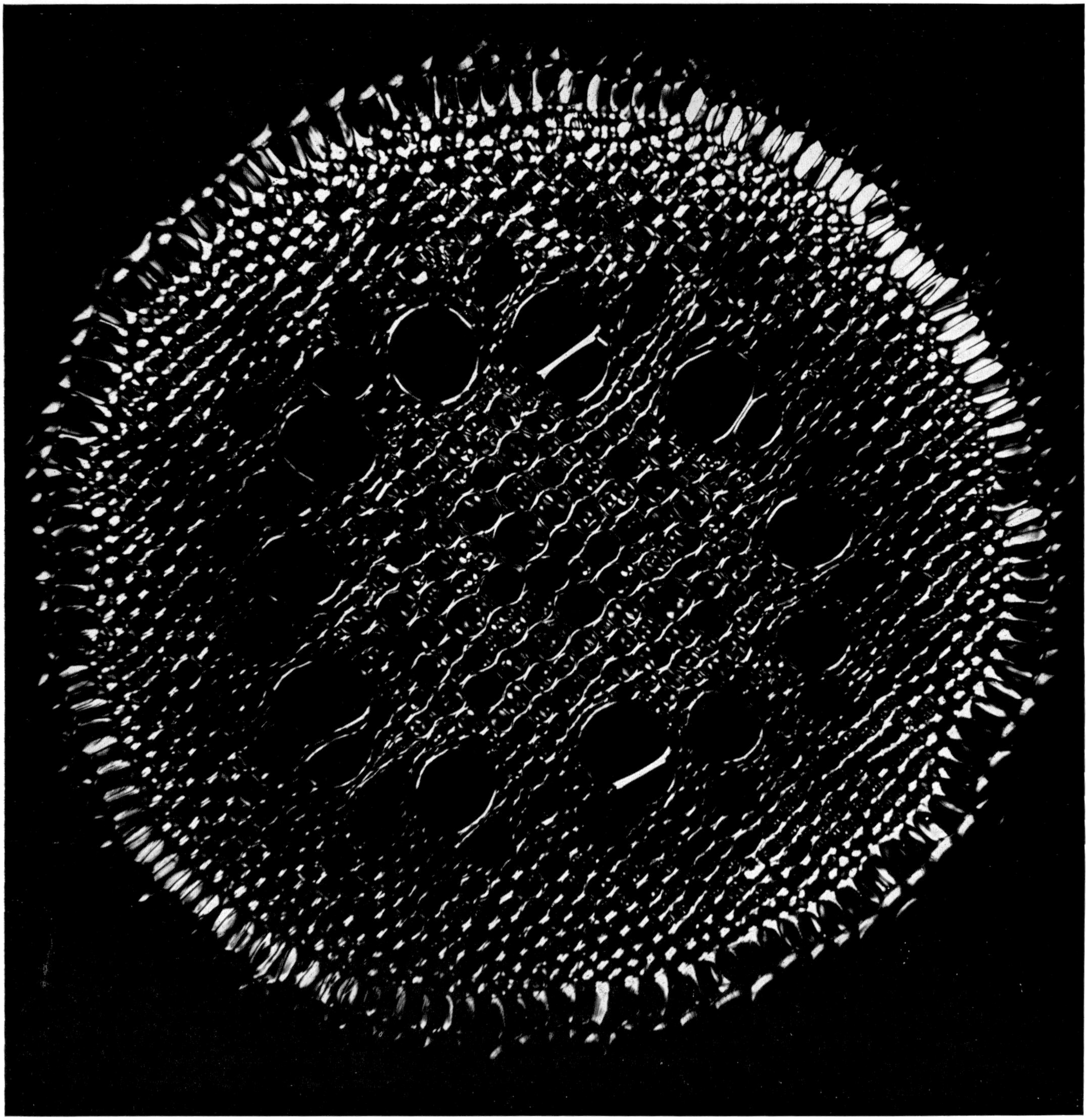

Frontispiece Transverse section of the stele of a young root of *Smilax* spp. (greenbriar) showing the heavily lignified walls of the peripheral endodermal cells. Polarized light. $\times$ 350.

Plant Structure and Development

A Pictorial and Physiological Approach

T. P. O'Brien

MONASH UNIVERSITY, AUSTRALIA

Margaret E. McCully

CARLETON UNIVERSITY, CANADA

The Macmillan Company / Collier-Macmillan, Limited, London

This book is dedicated to
Professor Ralph H. Wetmore

First Printing

Library of Congress catalog card number: 69-10318

The Macmillan Company

Collier-Macmillan Canada, Ltd., Toronto, Ontario

Printed in the United States of America

Foreword

It is a great honor to be asked to introduce this book to the public. I have known both the authors for some time, first as graduate students and then as colleagues, and have watched and shared their growing enthusiasm for the exploration of the fine structure of plant cells and tissues. Nevertheless, I have had no part whatever in the planning or preparation of this book and can therefore express an unconstrained opinion.

The first thing which will be noticed by the reader is the superb quality of the illustrations. Anatomical and cytological methods have developed rapidly within the last few years, and the authors have been responsible for one important improvement themselves. This is to prepare and fix sections for optical microscopy essentially like those for electron microscopy so that they can be cut very thin (though 5 to 10 times thicker than those for the electron microscope). The resulting high-quality fixation and minimal thickness results in sharpness and clarity of outline, as well as in contrast of staining, to an extent seldom seen in older preparations.

The second innovation, which is made evident from the title itself as well as every page of the text to the very end, is the close interrelation maintained between structure and function. This represents surely one of the great changes in emphasis and in organization which modern biology has undergone. At last the realization of evolution has permeated the minds and hearts of anatomists and physiologists, and we see *form* as it has evolved for the performance of function, not as a thing in itself. *Function* we see as having its qualities modified, and its limitations imposed, by form. Thus this book is an attempt to present plant structures, from whole organs to subcellular organelles and beyond, as functioning entities. Attention no longer centers solely on the mature form, but equally on the stages which lead to it; the structure is seen as one in the course of development, the cells as in the course of differentiation. A corollary to this approach is the considerable attention given to seedling material, which is commonly skimped in anatomical treatments. Seedlings are, of course, so widely used for physiological experiments that their prominence here further stresses the interrelation between structure and formation.

The book has perhaps a third quality, which by its very nature will *not* be noticed by the reader, at least at first. This is the fact that its deceptively simple and lucid style conceals a great depth of understanding, and indeed without such an understanding the presentation would inevitably have been more complex and turgid. As it is, a great mass of up-to-date information is presented in very succinct form. As I know at first hand, the authors have examined immense amounts of material to arrive at a close and yet not oversimplified description of each item. Also they know enough to recognize when the limits of present knowledge (anatomical or physiological) have been reached and only further research can give the desired explanation.

A book of this size must necessarily omit a great deal. The authors have clearly aimed at being selective rather than encyclopedic. But the curious reader may well be stimulated to fill some of the gaps by making *ad hoc* studies of his own, and thus the book could become like a whirlpool, trailing a chain of lesser whirlpools in its wake. The most influential books are those which have had such results.

The book should prove highly appropriate for modern-style courses on dynamic anatomy, or what Haberlandt long ago (with a vision which was unfortunately far ahead of its time) called *physiological anatomy*. More broadly, however, it should serve as an introduction to the structure of the living and developing plant for everyone interested in biology, from beginners upward.

KENNETH V. THIMANN
Crown College, Santa Cruz

Preface

THIS BOOK arose because of our experience as Teaching Fellows in several undergraduate courses in the Biological Laboratories at Harvard University. We noted that relatively few sophomores or juniors were interested in separate courses in plant physiology, plant anatomy, or plant morphology. On the other hand, when parts of the subject matter of these courses were presented in a more integrated manner as a course in developmental biology, student interest was high. In addition, we found that it was possible to teach plant anatomy and histology very effectively from good photomicrographs, supplemented with the barest minimum of time spent in drawing diagrams from prepared slides viewed with the microscope. These two facts guided our selection of material and the manner of its presentation. Most of the anatomy and histology is treated pictorially, and a positive attempt has been made to minimize description of the figures in the text. Where a long description seemed to be warranted, it has been incorporated in the figure legend. The text is an attempt to summarize what is known about the functions of the structures that are illustrated. In writing these summaries we have drawn freely upon information based in biochemistry, physiology, experimental morphology, and developmental biology.

Although the book offers an introduction to all of the organs of the higher plant, the depth of the coverage within each chapter is variable and incomplete. Nor could it be complete, for we did not wish to include topics of structural interest about whose functions little is known (for example, the embryo sac) nor topics of physiological interest whose structural basis is poorly understood (for example, the salt accumulation by roots). Thus we regard this book as an introductory and selective guide to the internal structures of higher plants and their functions. With this in mind, each chapter ends with a list of general references in addition to the references to the research and review papers indicated numerically in the text.

Several other factors influenced our presentation. We have often found that students have difficulty in relating light or electron micrographs of sections to the organ from which they were made. The book contains a number of photomicrographs, often placed at the beginning of a section, to help orient the student, and most of the electron micrographs are accompanied by a light micrograph of a similar specimen.

In an effort to discourage the use of nonspecific stains, we have limited ourselves, with few exceptions, to the use of staining procedures for which there is a firm histochemical basis. Each figure legend includes the stain used. The details of the procedures, together with the methods of specimen preparation for light microscopy, are contained in the Appendix.

In choosing the vocabulary of this book we have been influenced by the upsurge of biology courses, as distinct from botany, zoology, and microbiology courses. Because there are fewer botanists than zoologists or microbiologists, it seems likely that these courses will come to be taught increasingly by staff who lack a thorough grounding in botanical vocabulary. Accordingly, we have eliminated specialized terms wherever possible, and when faced with a choice between two terms, we have selected the more descriptive one (for example, *cell membrane* instead of *plasmalemma, vacuolar membrane* instead of *tonoplast*). For similar reasons, mature tracheary elements that consist only of a cell wall are not called cells, and we have substituted *provascular tissue* for the more confusing term, *procambium.* Wherever a new term is introduced for the first time, it is italicized.

Although our thanks are due to many people for their help in preparing this book, to no one are we more indebted than to Dr. Ned Feder. He introduced us to the teaching of histology from photomicrographs and he instructed us initially in the use of the fixation, embedding, and staining procedures upon which the photomicrographs of this book are based. Our gratitude is very deep. We are also grateful to the following, who provided us either with specimens for photomicrography or with photomicrographs: Dr. A. Bajer, Mr. J. Brown, Dr. F. A. L. Clowes, Dr. R. N. Govier, Mr. A. D. Greenwood, Dr. B. E. S. Gunning, Dr. B. E. Juniper, Dr. M. C. Ledbetter, Dr. E. H. Newcomb, Dr. J. S. Pate, Mr. J. R. Pilcher, and Dr. D. S. Skene. The following students of Carleton University also supplied us with material for illustrations: Mr. J. Bell, Mr. F. T. Bellware, Mr. K. R. Brasch, Mrs. M. Brasch, Mr. R. G. Fulcher, Mr. G. Glantz, Miss J. Millar, Mr. R. D. Muir, and Mr. H. B. Younghusband. We are indebted to Professor D. J. Carr of the Queen's University, Belfast, for placing the facilities of the Botany Department at our disposal while we completed the photomicrography, to Mrs. D. J. Carr for her invaluable assistance with the photomicrography, and to Dr. B. E. S. Gunning for his time, tolerance, and help with the photomicrography.

One of us (T. P. O'B.) was a member of the Society of Fellows of Harvard University during the period of preparation of this book. We wish to acknowledge the financial support of the Society which made it possible to collect specimens, test methods of preparation, and to travel to Belfast to complete the book.

We would also like to acknowledge a grant from Carleton University and a visitor's fellowship from the British Science Research Council which made it possible for one of us (M. E. McC.) to work on the book at the Queen's University, Belfast.

T. P. O'B.
M. E. McC.

Contents

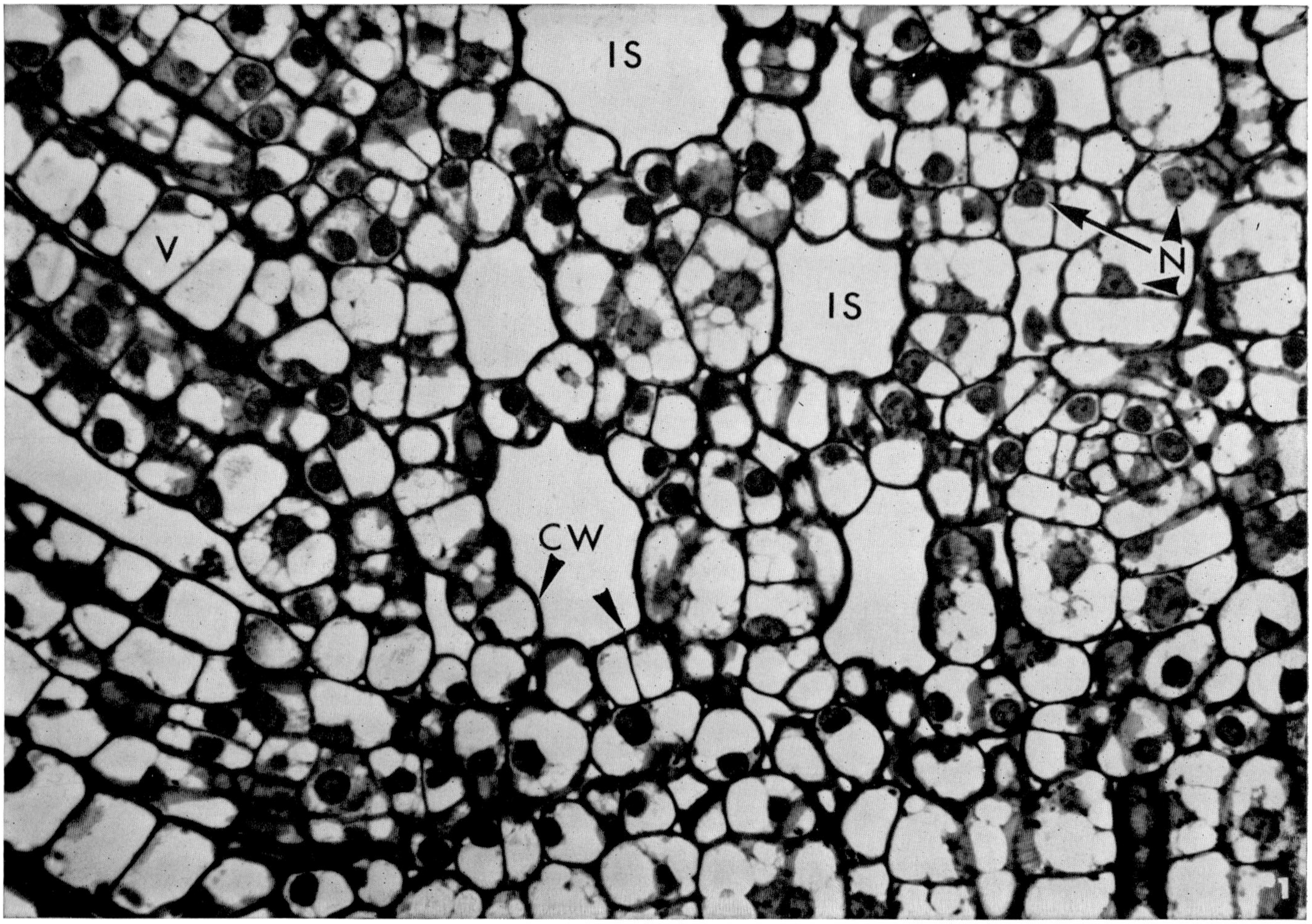
IS
IS
V
N
CW
1

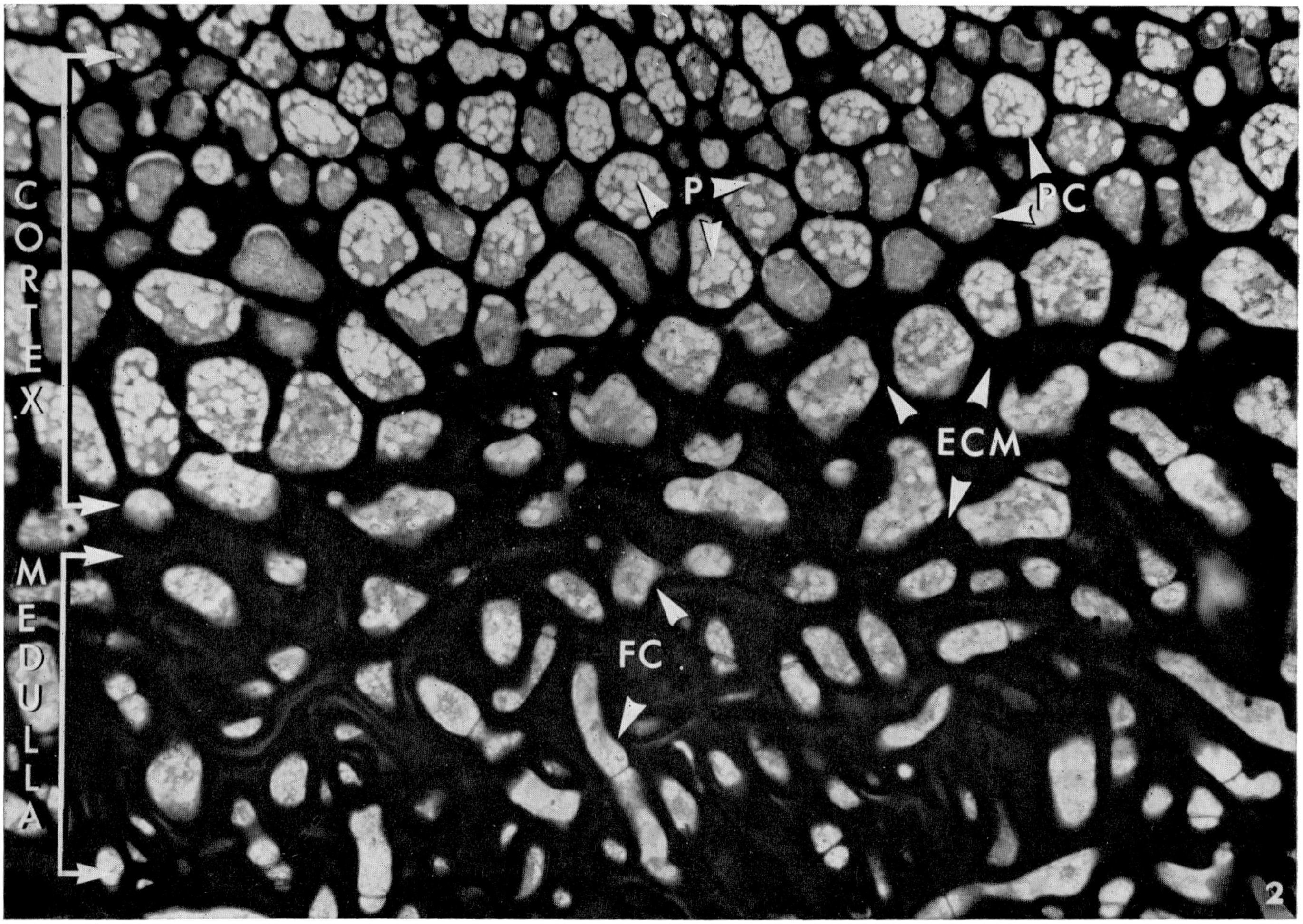
CORTEX
MEDULLA
P
PC
ECM
FC
2

Introduction

THE GENERAL appearance of the major organs of higher plants—roots, stems, leaves, buds, flowers, and fruits—is familiar to all. If one cuts any of these organs into thin slices (10–25μ thick; 1μ equals 10^{-3} mm) with a razor blade and examines them with a microscope, it is easy to see that organs consist of *cells* and *extracellular material.* The ratio between these two components, their distribution within the tissue, and the proportion of tissue volume occupied by gas spaces vary widely from organ to organ. In general the extracellular material is present as a discrete layer—the cell wall—which surrounds each cell (Figure 1). The chemical composition and physical properties of the wall vary sharply with cells of different function: indeed, variation in these properties is the basis of classification of most types of plant cells. In many instances, however, and especially in the marine algae, the extracellular material is not present as a clear-cut wall surrounding each cell (Figure 2). In such cases, it becomes impossible to decide which cell has elaborated certain parts of the extracellular material. Despite these important exceptions, it is helpful to regard the cell and its associated extracellular material—the cell wall—as the unit of structure, and we shall begin to examine this unit in considerable detail.

Many of the illustrations in this book are light or electron micrographs, and it is important to remember that the living specimens from which they have been prepared were fixed, dehydrated, embedded, sectioned, and stained prior to photography. All of the water, salts, and organic compounds of low molecular weight have been lost from the tissue; only the macromolecules are retained. Even with the best preparation procedures, some changes in the architecture of this macromolecular framework must have occurred. A brief look at living cells under phase contrast should convince the most optimistic that the relationship between the static image and the dynamic reality is at best uncertain. One must, therefore, exercise great caution in generalizing from structure preserved

1 **Longitudinal section of a stem of *Elodea densa.* The tissues consist chiefly of parenchyma cells around each of which the extracellular material is organized as a discrete cell wall (CW). The nuclei (N) are distinctive and the cells contain large vacuoles (V). Note the intercelluar gas spaces (IS), some of which are very large in this aquatic plant. The bases of several immature leaves are seen on the left side of the photomicrograph. Periodic acid/Schiff's reaction (PAS)/toluidine blue. × 640.**

2 **Transverse section of the thallus of *Fucus vesiculosus,* a marine brown alga. Contrast the tissue organization here with that illustrated in Figure 1. There are no gas spaces, and although the parenchyma cells of the upper cortex (PC) have more or less discrete cell walls, those of the middle cortex and medulla do not. Indeed the filament cells (FC) in the medulla are embedded in extracellular material (ECM), much of which they did not produce. The parenchyma cells contain an abundance of plastids (P). Toluidine blue. × 480.**

at a moment in time, and revealed in these photographs, to that which must have obtained in the living state.

Finally, two comments on interpretation. In the text we have often found it convenient to describe the three-dimensional appearance of structures (as they exist *in vivo*) even though they only appear in two dimensions in the sections illustrated. Secondly, most of the sections shown are very thin (0.5–3μ for light micrographs, 0.05–0.1μ for electron micrographs). Many cells, therefore, appear to lack nuclei because only a slice of each cell is shown.

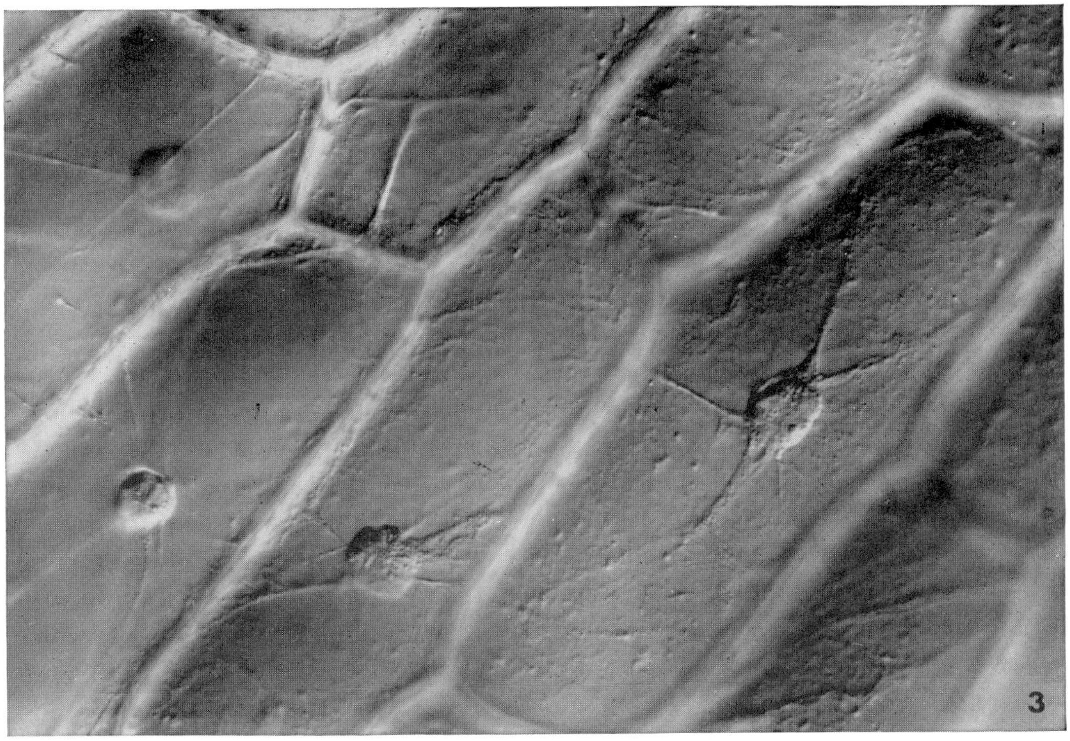

3 Interference contrast (Nomarski optics) photomicrograph of living cells of the lower epidermis of a bulb scale of onion (*Allium* spp.). In each cell the nucleus lies at the focus of numerous fine strands of cytoplasm which traverse the vacuole. × 200.

1
The Cell

ALL CELLS arise by the growth and division of pre-existing cells. Zones of active cell production are termed *meristems,* and *meristematic cells* from the tip of a growing root (see Figures 4 and 5) are a convenient cell type with which to introduce subcellular organization.

What one sees in sections of suitably fixed root-tip cells depends to a certain extent upon how the section is stained and the optical system employed (see Appendix). Figure 4 shows the appearance of cells stained with iodine and examined by phase contrast. Each cell contains a large, approximately spherical body, the *nucleus.* Within it lies a densely stained *nucleolus,* which may contain one or more *nucleolar vacuoles.* A thin line marks the position of the *nuclear envelope,* the boundary that separates the *nucleoplasm* from the *cytoplasm.* The cytoplasm consists of a parietal layer and numerous fine strands that traverse the *vacuoles,* sap-filled cavities that are separated from the cytoplasm by the *vacuolar membranes* (or *tonoplasts*) (see also Figure 15 and the living epidermal cells in Figure 3). A population of small organelles, consisting of small spheres, plump rods, filaments, and granules can be distinguished within the cytoplasm. This population includes *plastids, mitochondria,* and *dictyosomes.* Unfortunately, it is not possible to distinguish with certainty between these organelles by phase contrast microscopy, but the plump rods and larger spheres are probably plastids and the finer filaments are probably mitochondria.

Figure 5 shows the same field as Figure 4 stained with the cationic dye toluidine blue O, which reacts with negatively charged groups ($H_2PO_4^-$, HSO_4^-, COO^-). *Chromatin* is distributed throughout the nucleus as aggregates of different size. The cytoplasm and nucleolus are also strongly stained. The affinity of these components for cationic dyes (loosely termed *basophilia*) is due largely to the phosphate groups ($H_2PO_4^-$) of deoxyribonucleic acid (DNA) in the chromatin and to the phosphate groups of ribonucleic acid (RNA) in the nucleolus, nucleoplasm, and cytoplasm. The cell walls are also strongly stained, owing to the presence of carboxyl groups (COO^-) in polyuronides. The discontinuous appearance of the cell wall is due to the presence of numerous *primary pit fields,* thin areas of the wall through which the *cell membranes* of adjacent cells are in contact at minute structures, the *plasmodesmata* (see also Figures 7 and 8).

Structures that lie at the limit of resolution of the light microscope may be examined in detail in the electron microscope. Figure 6 is an electron micrograph showing parts of two cells from a root tip that was fixed in glutaraldehyde and treated with osmium tetroxide (OsO_4) (*1.1*). OsO_4 reacts with a variety of chemical groups in the tissue (*1.2*). The electron contrast caused by the presence of this heavy metal in the tissue has been further enhanced by "staining" the sections with uranyl acetate (*1.3*) and lead citrate (*1.4*) prior to photography in the electron microscope. Unfortunately, the chemistry which underlies the reactions of OsO_4 with aldehyde-fixed tissue components and the further reactions of OsO_4-treated tissue with these heavy-metal salts are largely obscure. Despite these limitations a great deal may be learned about subcellular structure from such specimens.

In this particular tissue most of the chromatin is finely dispersed throughout the nucleoplasm, although a few aggregates are evident at the inner margins of the nuclear envelope and near the nucleolus. The nucleolus shows evidence of both granular and fibrous components (see also Figures 16 and 17). The nuclear envelope appears as two dark lines that bound a space of more or less constant width. The cytoplasm and nucleoplasm appear to be in contact through this envelope at numerous sites, the *nuclear pores* (solid arrows, Figure 6). Mitochondria (readily distinguished from the plastids by the arrangement of their internal membranes) and several dictyosomes can be seen.

Small, dark granules, the *ribosomes,* either lie free in the *ground substance* of the cytoplasm or occur bound to the surface of the *endoplasmic reticulum.* Endoplasmic reticulum, when covered with ribosomes, is commonly called "rough" endoplasmic reticulum. The basophilia of the cytoplasm in Figure 5 is due to the presence of numerous ribosomes, each of which contains RNA.

The endoplasmic reticulum consists commonly of *cisternae,* flattened sacs bounded by a membrane. The cisternae are inter-connected and when viewed in profile resemble the nuclear envelope (*1.5*). The cytoplasmic side of the nuclear envelope often has ribosomes bound to it. It has been demonstrated in many cell types that the nuclear envelope is connected to the endoplasmic reticulum and may be considered to be a special part of this reticulum. The morphology of the endoplasmic reticulum varies a great deal in cells of different function and ribosomes do not always occur on its membranes.

It may also vary in different locations within the cell. A tubular form, filled with a material rich in protein, is present in the lower right hand corner of Figure 6.

Populations of vesicles, each bounded by a membrane, occur throughout the ground substance and are especially evident in the neighborhood of the dictyosomes (*1.6*). Some of these vesicles are illustrated in more detail in Figures 11 and 12. The ground substance itself contains many of the soluble enzyme systems of the cell but appears in electron micrographs merely as a mat of fine fibrils of rather low electron density (*1.7*).

Each cell is surrounded by a cell wall (Figure 6 and inset), which often shows granular or fibrous substructure. To date, it has not been possible to establish just which components of the cell wall are responsible for this appearance (*1.8*). The cytoplasm is bounded by the cell membrane (also termed the *plasmalemma*). Since the thickness of the wall varies in an irregular fashion, the cell membrane, which is closely applied to the wall, follows a tortuous path. The cell membrane invariably stains more strongly on the side facing the wall, a fact that may be connected with the synthesis of certain wall components at the cell surface (*1.9*).

The interpretation of the structure of plasmodesmata is highly controversial (*1.10, 1.11, 1.12*). Most workers agree that each plasmodesma consists of a membrane-lined canal and that this membrane is continuous with the cell membranes of the two adjacent cells. The question of continuity of the endoplasmic reticulum through these canals lies at the center of the controversy. In our view, the endoplasmic reticulum comes to and touches a plug of electron-dense material which closes the neck of each canal at the cytoplasm surface (Figure 7). We do *not* feel that the central core, which appears to be continuous with the closing plug, and which occupies part of the lumen of the canal (solid white arrows, Figures 7 and 8), is endoplasmic reticulum. However, the endoplasmic reticulum does penetrate the young cell plate during its formation at the close of cell division (*1.13*) (see Figures 35 and 36).

The central core is about 250Å in diameter and in oblique sections at various depths in the wall (Figure 8) shows no evidence of membrane structure. Clearly these minute structures, unique to plants, need more careful examination, for it is imperative to know whether they really constitute a preferential pathway for the movement of large molecules from cell to cell.

The structures (annular when seen in cross section) that lie just beneath the cell membrane in the inset to Figure 6 were first recognized in higher plant cells just a few years ago (*1.14*). Grazing sections of the wall–cytoplasm boundary (Figure 9) reveal that these *microtubules* are cylindrical structures 250–270Å in diameter and of indefinite length. The discovery of these microtubules in the cell cortex has stimulated a great deal of interest and controversy. Many investigators have found that their orientation is parallel to that of the cellulose microfibrils in the most recently formed layer of the wall (*1.14, 1.15, 1.16, 1.17*). In addition, they are morphologically similar to the microtubules that are present in spindle fibers (see Figures 25 and 27) (*1.14*). Just prior to nuclear division, they form a band the position of which reflects quite accurately the plane of the future cell plate in both symmetric and asymmetric divisions of the cell (*1.18*). Drugs such as colchicine, which appear to disorganize microtubules, also drastically modify the pattern of wall deposition, the movement of chromosomes, and the development of the cell plate (*1.19*).

The wall of each microtubule seems to be composed of filaments, some evidence of which is just discernible in the inset to Figure 6. Several groups of workers are currently engaged in attempts to isolate microtubules and characterize their proteins. It is already clear that our knowledge of these structures, whose distribution is correlated with so many important aspects of cell function, is only in its infancy.

Mitochondria

When mitochondria (*1.20, 1.21*) are examined in the living cell, it is often easy to see that their size and shape (spheres, dumbbells, or elongated filaments which are often branched) can alter markedly within a few minutes. Unfortunately, one does not gain an adequate impression of this dynamic state of mitochondria from an examination of sectioned material. In thin section, the structure of mitochondria from different sources is remarkably constant. Each mitochondrion invariably has two distinct membrane systems, an outer one that abuts the ground substance of the

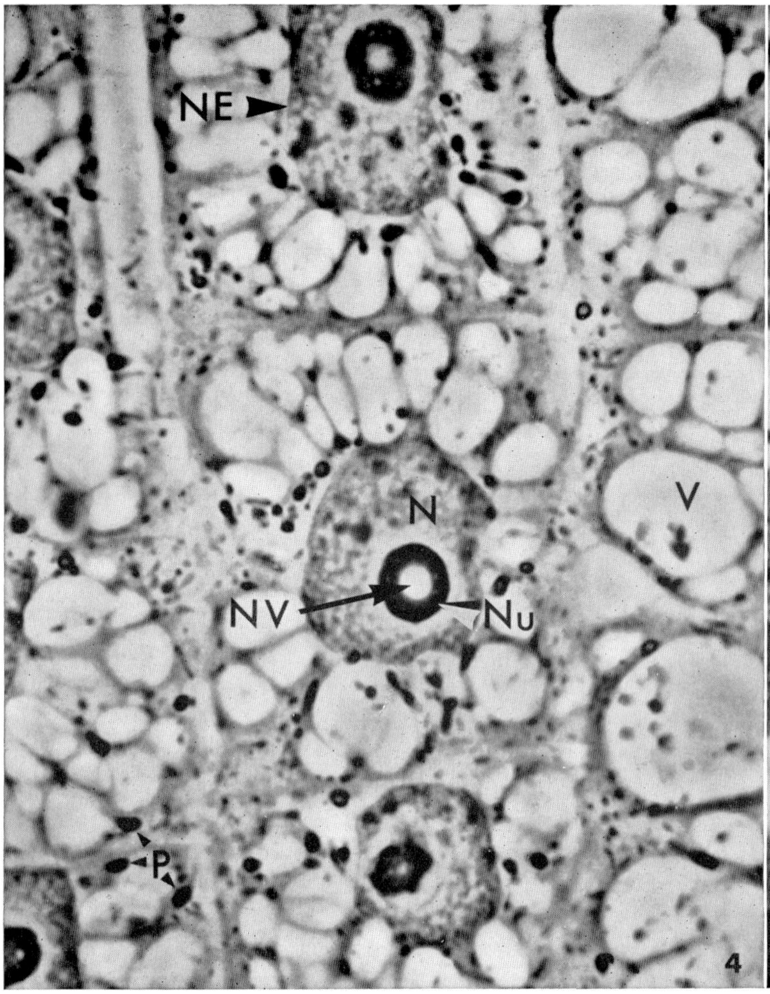

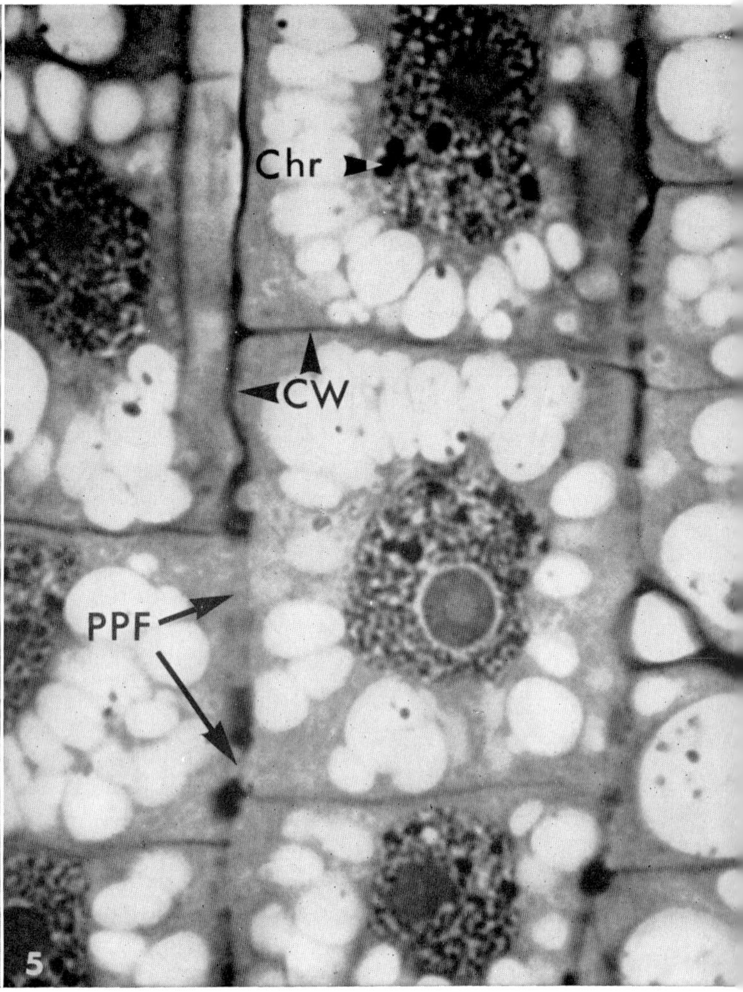

4 Interphase cells from the root tip of *Vicia faba* (broad bean), stained with iodine and photographed under phase contrast. N, nucleus; NE, nuclear envelope; Nu, nucleolus; NV, nucleolar vacuole; P, plastid; V, vacuole. × 2100.

5 Same section as that shown in Figure 4, stained with toluidine blue. Chr, chromatin; CW, cell wall; PPF, primary pit field. × 2100. (Section courtesy Dr. B. E. S. Gunning.)

cytoplasm, and an inner one whose invaginations, the *cristae mitochondriales*, give to the organelle its characteristic appearance in thin sections. The morphology of the cristae, the number of cristae per mitochondrion, and the number of mitochondria per cell all vary significantly among cells of different function. In general, cells with high rates of aerobic respiration may show an increase in mitochondrial numbers, increase in cristae per mitochondrion, or both relative to cells with more modest levels of oxygen consumption and energy expenditure. Because the cristae are invaginations of the inner membrane (Figure 10, black arrow), the space enclosed by the cristae is continuous with the space between the two membranes. The inner membrane also encloses a space of complex

6 Electron micrograph showing part of two cells from the root tip of *Arabidopsis thaliana* (mouse-ear cress). Note nucleus, with prominent nucleolus (Nu), nuclear envelope (NE) with pores (arrows), plastid (P), mitochondria (M), dictyosomes (D), and rough endoplasmic reticulum (ER). × 20,000. *Inset:* Cell wall (CW), cell membrane (CM), and transverse section of cortical microtubules in root-tip cell of *Phleum pratense* (timothy). × 120,000. (Both courtesy Dr. M. C. Ledbetter.)

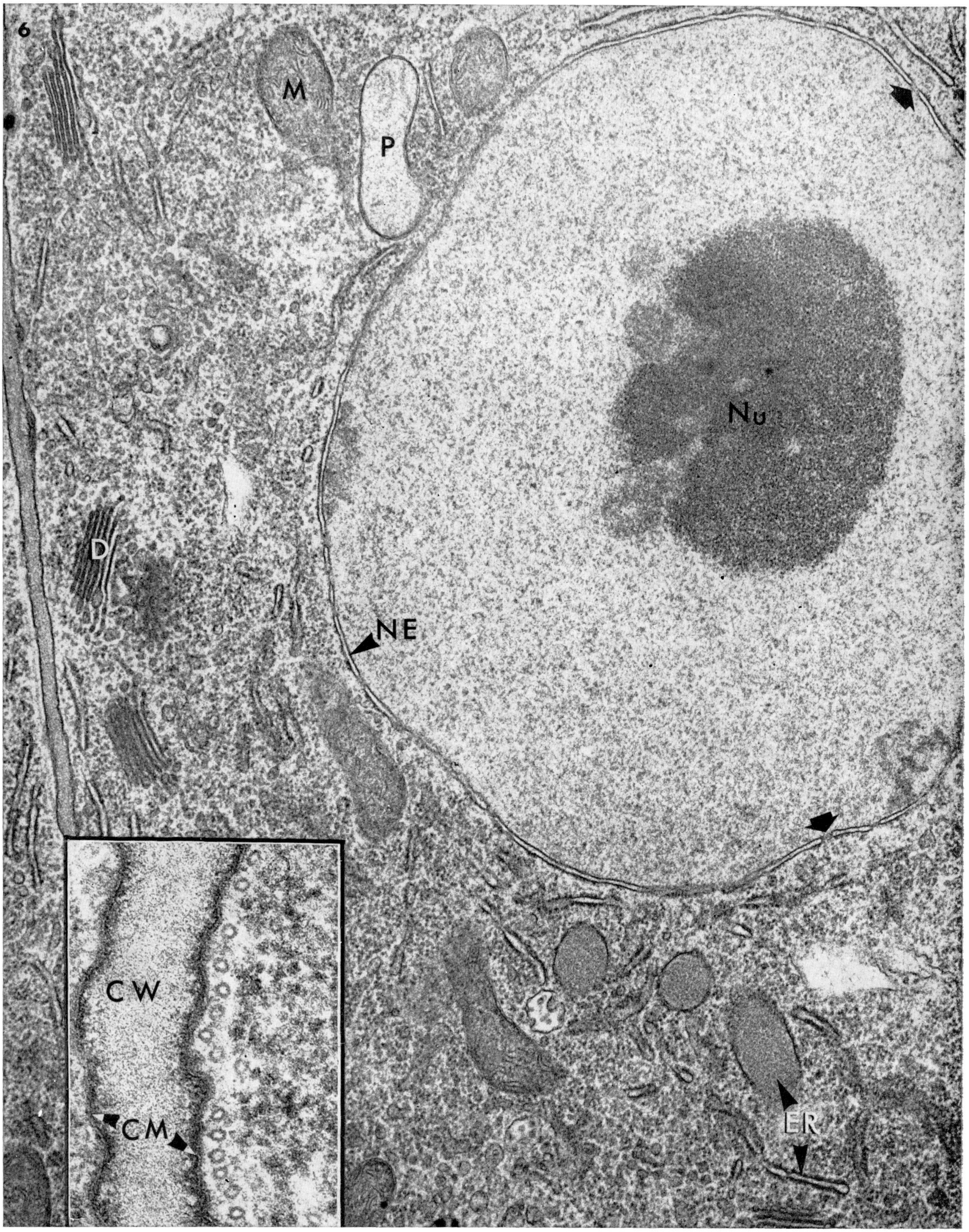

6
M
P
Nu
D
NE
CW
CM
ER

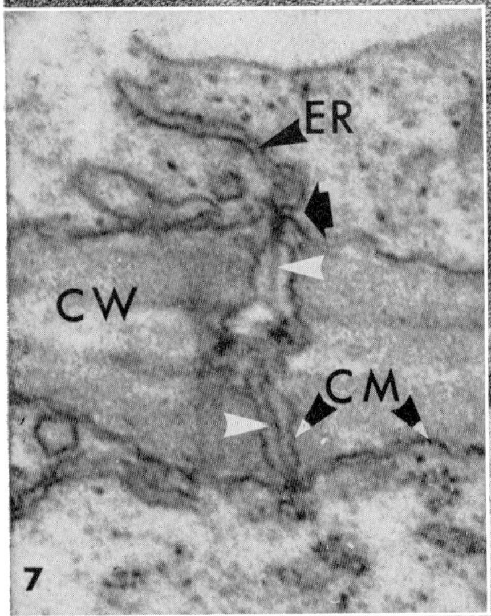

7

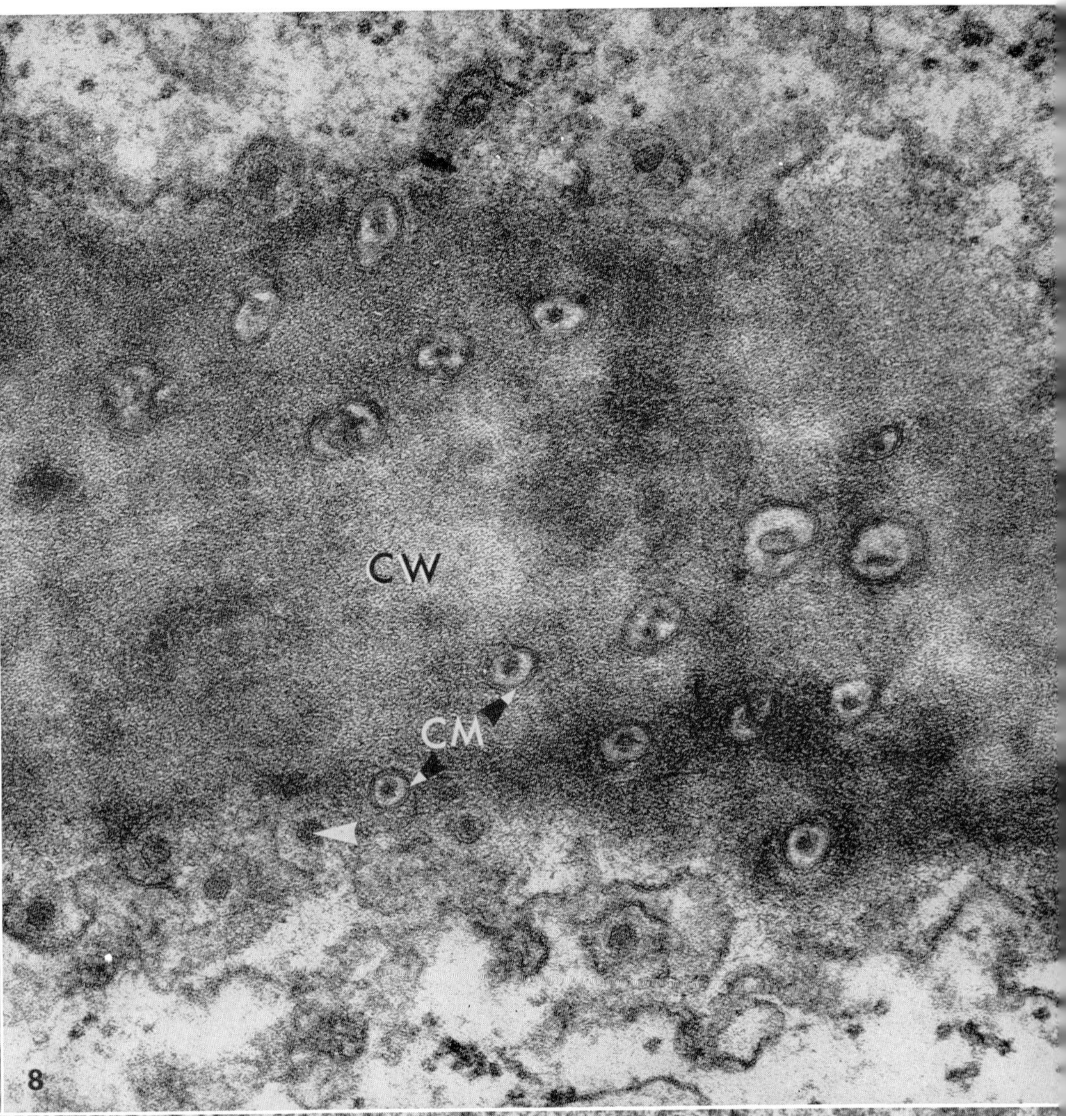

8

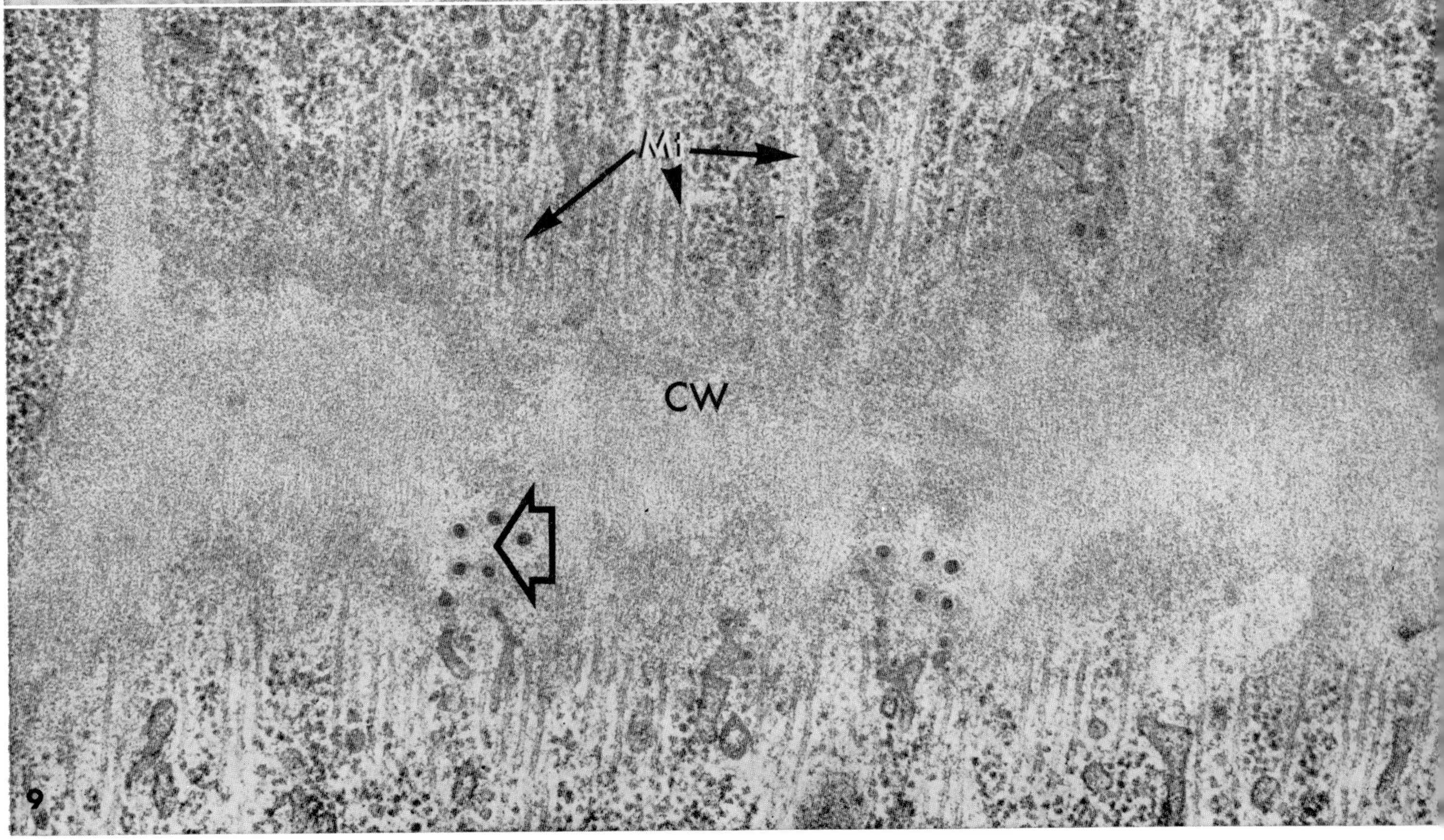

9

shape filled with the *mitochondrial matrix* within which lie one or more *osmiophilic granules.*

This remarkably simple, constant morphology gives only the merest inkling of the complexity of functions carried out by these organelles. It has been known for some time that the enzymes of the Krebs citric acid cycle, and the enzymes responsible for oxidation of fatty acids and trioses, along with the electron transport pathway that is coupled to phosphorylation, are all located within mitochondria. In actively respiring tissues, these activities lead to the synthesis of ATP, which is required as a source of chemical energy in most synthetic reactions of the cell. For this reason, the mitochondria are often called, quite aptly, "the power houses" of the cell. Recently, it has been shown that mitochondria can accumulate calcium phosphate during oxidative phosphorylation and that the osmiophilic granules referred to above are the sites at which the accumulated phosphate is precipitated (*1.22*).

The assemblies of respiratory enzymes upon which so many of these mitochondrial properties depend are located in the membrane fraction of isolated mitochondria. These assemblies account for as much as 25 per cent of the membrane protein, and it is commonly held that they are a structurally significant fraction of the inner membrane and cristae. Other more readily extracted enzyme systems are believed to be present in the matrix.

A property of mitochondria that is as yet poorly understood and that may underlie the changes in shape seen *in vivo* is their capacity to undergo reversible swelling and contraction *in vitro,* with variation in volume of as much as 100 per cent (*1.23*).

Finally, most workers believe that in growing cells, new mitochondria arise by the growth and division of pre-existing mitochondria (*1.24*). It has been shown that mitochondria can synthesize some, but not all, of their lipids and can incorporate labelled amino acids into membrane protein. Recently it has also been shown that these organelles contain their own somewhat distinctive DNA (*1.25, 1.26*). These facts, coupled with mitochondrial size, oxidative activity, and capacity for ion transport, have led several workers to propose quite seriously that mitochondria are highly evolved intracellular symbionts, derived perhaps from organisms resembling modern bacteria.

Dictyosomes

These organelles (*1.27*) can be distinguished with certainty in plant cells by light microscopy only in the most favorable circumstances (*1.28*) and knowledge of their structure depends almost entirely upon electron microscopy. A dictyosome consists of a stack of cisternae each of which is more or less entire and somewhat flattened in the center of the stack. At the margins the cisternae are fenestrated, and in dictyosomes isolated from the cell, the edges consist of a network of anastomosing tubules. In the central part of the stack, fine filaments, about 70Å in diameter and 150A

7 **Longitudinal section of plasmodesmata in the parenchyma cell walls (CW) of a coleoptile of oat (*Avena sativa*). Each plasmodesma consists of a canal lined with a membrane that is continuous with the cell membrane (CM; white-tipped black arrows). A core of dense material (white arrows) lies within this canal and is continuous with a plug that closes the canal at the cell surface. The endoplasmic reticulum (ER) appears to touch these closing plugs (large black arrow). × 75,000.**

8 **The appearance of plasmodesmata in cells similar to those shown in Figure 7. The wall (CW) is sectioned at an angle exposing the plasmodesmata in profile at different levels in the wall. The cell membrane (CM), which lines the canal (white-tipped black arrows), and the central cores are readily apparent. × 96,000. (Both reproduced with permission from *Protoplasma,* 63, 1967, pp. 426–429).**

9 **Grazing section of wall/cytoplasm boundary in cells from the root tip of *Arabidopsis thaliana,* showing microtubules (Mt) in surface view (compare with Figure 6, inset). The open arrow shows a primary pit field (PPF) with plasmodesmata (see also Figure 5). × 57,000. (Courtesy Dr. M. C. Ledbetter.)**

apart, lie between the cisternae. The function of these filaments is uncertain.

Many workers believe that each dictyosome is a polarized structure with a forming face at which new cisternae arise and a mature face at which *dictyosome vesicles* are produced and released into the ground substance of the cell. If this view is correct, a profile image of a dictyosome (as in Figure 11) is an array in space of the developmental sequence through which each cisterna will pass in time.

Material whose electron contrast often closely resembles that of the cell wall is commonly found both in the dictyosome vesicles and in the more mature cisternae (Figure 12, asterisks). Experiments with tritiated glucose (glucose labelled with the radioactive isotope tritium—^3H) and electron-microscope autoradiography have confirmed beyond reasonable doubt earlier suggestions that the dictyosomes are involved in the production of noncellulosic polysaccharides (*1.29*). These polysaccharides are incorporated into the wall by fusion of the dictyosome vesicles with the cell membrane.

Simple calculations suggest that this process of vesicle incorporation at the cell surface must supply an enormous area of membrane to the cell membrane, an area that, in rapidly elongating cells, could be as much as ten times the surface available for it. How this membrane excess is adjusted is not known: it could be destroyed, or it could be recovered from the surface by the formation of pinocytotic vesicles. In either case it is clear that the cell membrane must be regarded as a very dynamic entity.

10 **Mitochondria in a xylem parenchyma cell of an oat coleoptile. The cristae (Cr) clearly arise as invaginations of the inner of the two membranes that bound these organelles (black arrow). Note the osmiophilic granules (white arrows), which are believed to be the sites at which calcium phosphate is deposited. × 48,300.**

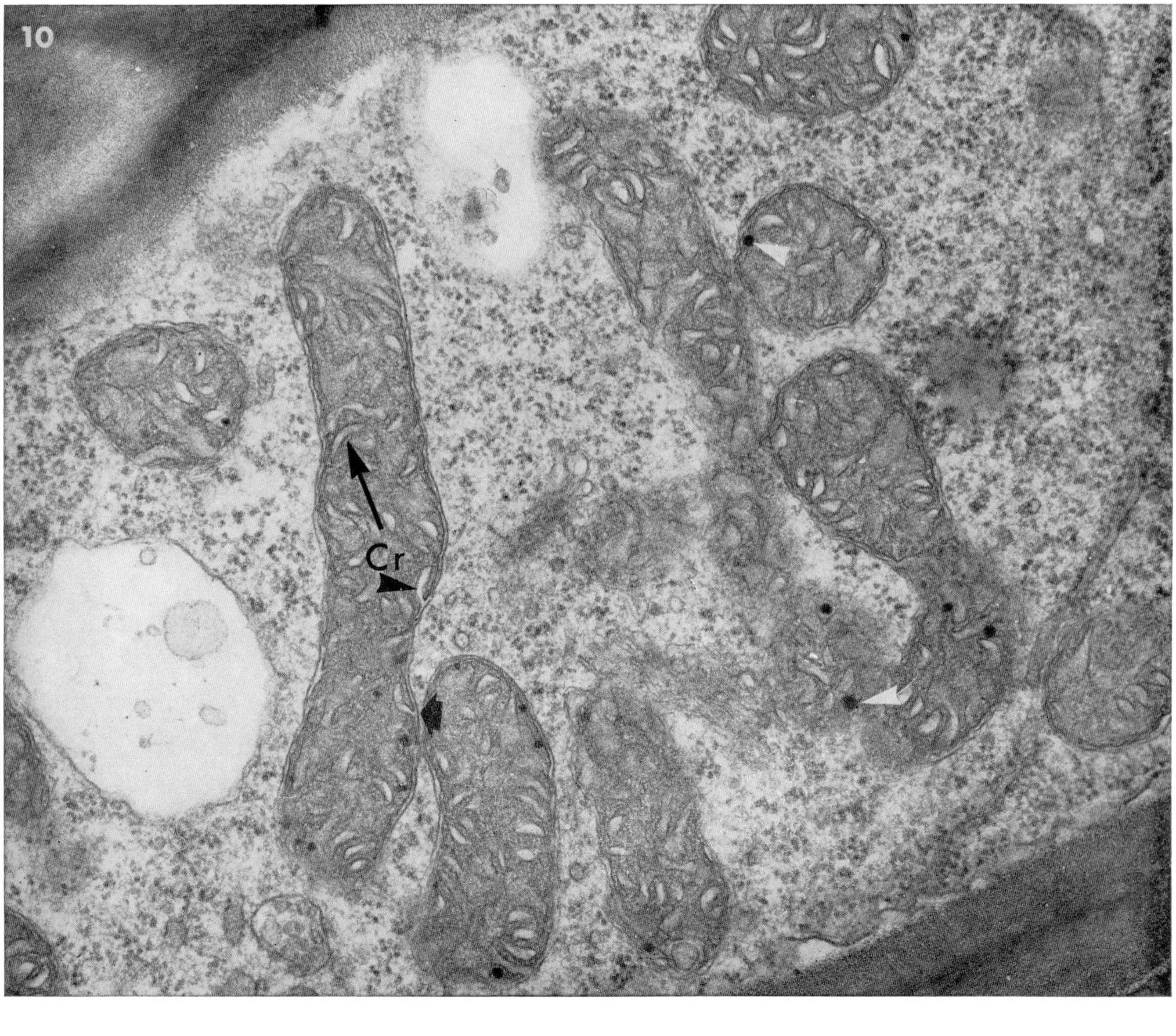

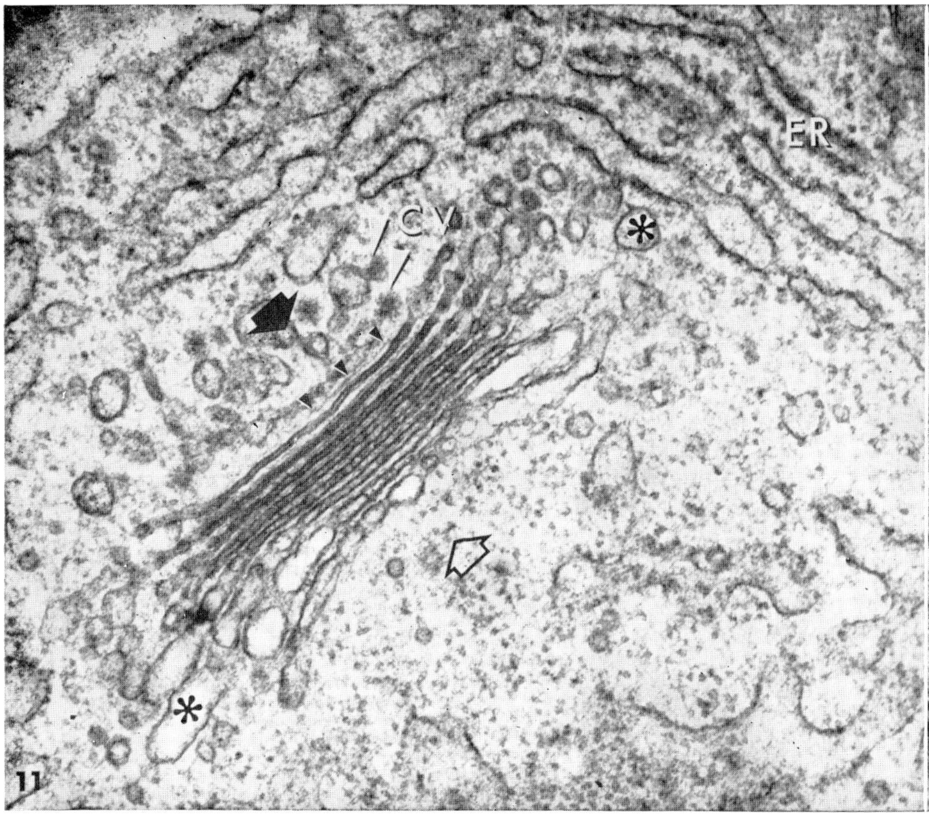

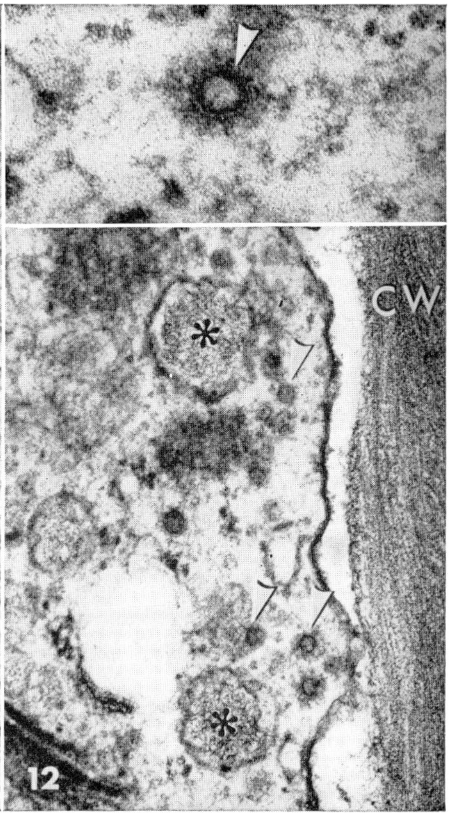

11 **A dictyosome in a cell of a glandular hair from a leaf of *Phaseolus vulgaris* (kidney-bean). Each dictyosome is polarized with a forming face (solid black arrow) and a secretory face (open black arrow), from the margins of which the dictyosome vesicles (asterisks) are released. Note the coated vesicles (CV) at the forming face and the intercisternal elements (row of small arrows) between adjacent cisternae. ER, endoplasmic reticulum. × 70,000.**

12 **Dictyosome vesicles (asterisks) and coated vesicles (arrows) in a cell of the outer epidermis of an oat coleoptile. Note that the content of the dictyosome vesicles resembles the nearby wall (CW) in staining properties. × 72,000. *Inset:* A coated vesicle at high magnification. × 150,000.**

Another type of vesicle, the *coated vesicle* (also known variously as shaggy, hairy, or alveolate vesicles), has also been identified both at the cell surface and attached to the cisternae on the forming face of the dictyosome (*1.27*). Vesicles with similar morphology have been shown to play a role in protein uptake from the cell surface in a variety of animal cells (*1.30*). In the plant cell it is not yet possible to decide in which direction these vesicles are moved, for one cannot distinguish between fusion and elaboration. Most workers have favored the view that these coated vesicles are secreting material to the surface in the plant cell (*1.6*), but it is also possible that they represent stages in the recovery of membrane (and perhaps enzymes associated with it) from the cell surface.

The number of dictyosomes per cell seems to vary markedly in cells of different function; estimates range from 30 in a meristematic root-tip cell to 30,000 in the large rhizoid cells of some algae. In addition, the number of cisternae per dictyosome, the size of the cisternae, and the number of dictyosome vesicles associated with each dictyosome can vary with the age of the cell and its state of activity. In general, cells engaged in the deposition of wall material tend to show hypertrophy of these features.

At certain times in the life of the cell (for example, at cell-plate formation; see Figure 35) the dictyosomes appear to occupy a special area in the cell. Whether they maintain stable positions in the cell at all times is not known but it seems somewhat unlikely since cytoplasmic streaming would probably cause them to move. Nevertheless, the forming face of dictyosomes is often in close proximity to cisternae of the endoplasmic reticulum

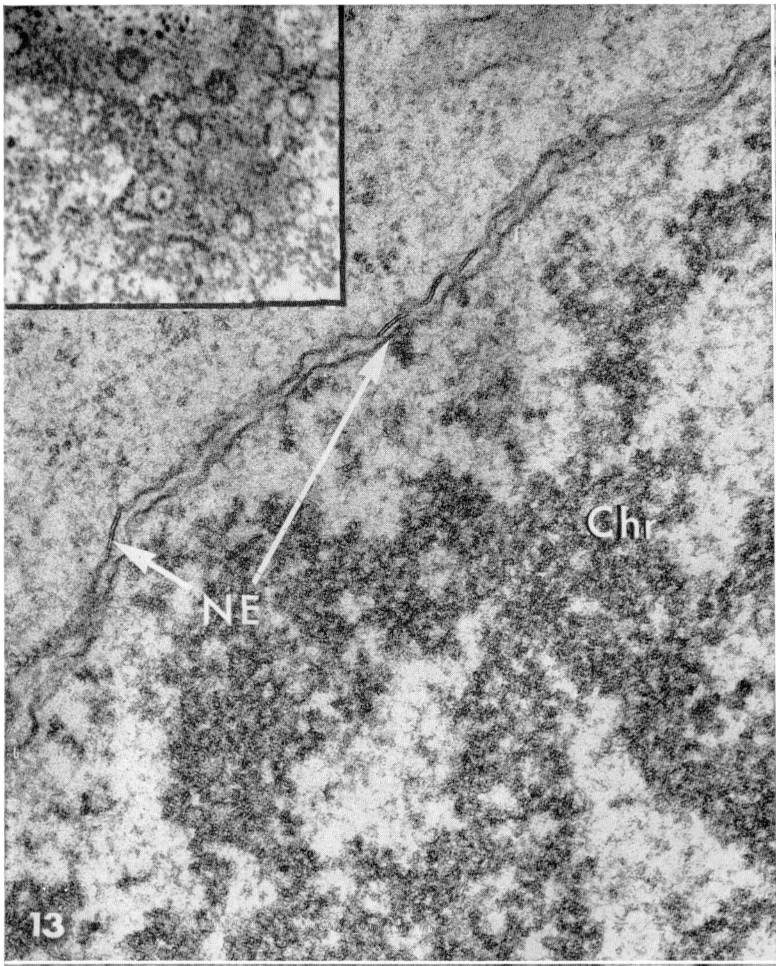
NE
Chr
13

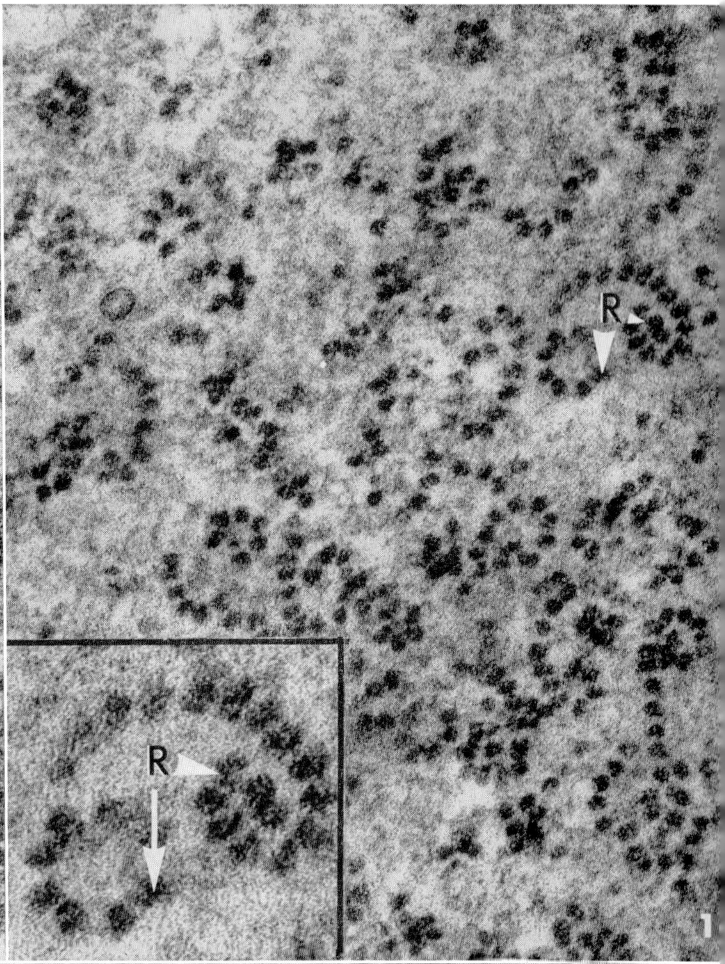
R
R
1

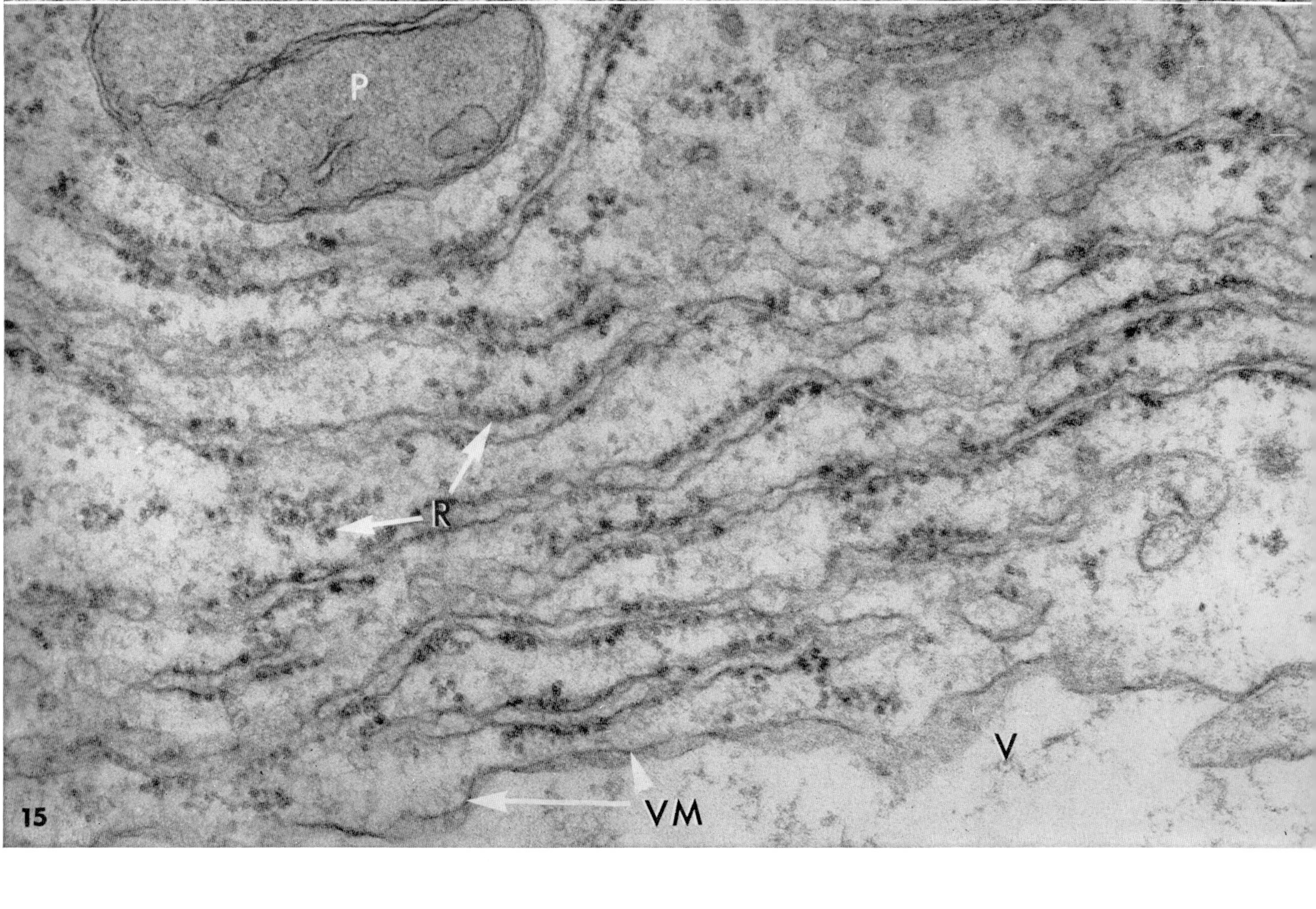
P
R
V
VM
15

(Figure 11). The stability of this association is questionable but the fact of its existence is suggestive in view of work in animal cells demonstrating the transfer of material from endoplasmic reticulum to dictyosomes. Until dictyosomes can be identified reliably *in vivo*, this important question cannot be answered.

The Nucleus and Nuclear Envelope; the Endoplasmic Reticulum and Ribosomes

Shortly after the turn of this century, the *chromosome* was shown to be the structural basis of heredity. Many aspects of this field of biology have advanced at an extraordinarily rapid rate in the past fifteen years. DNA is now recognized as the primary form of information storage of all cells and RNA as the intermediary through which this information is translated into specific protein synthesis. This subject and its history are treated in detail in most elementary texts and this information is not duplicated here. Rather, we wish to focus attention on some of the structures concerned with these activities.

Although the amount and distribution of the nuclear chromatin in the interphase cell often varies a great deal among different cell types, the chromatin pattern is often remarkably constant for any one cell type of the same organism. Some elegant experiments using autoradiography have shown that RNA synthesis is confined to the nucleus (*1.31*) (though there is evidence that other DNA-containing organelles, such as plastids and mitochondria, can also synthesize RNA) and that this activity is sharply repressed when the chromatin condenses during mitosis to form chromosomes (*1.32*) (see also Figures 19–23). These findings led to the idea that chromatin, a condensed DNA-protein complex, is relatively inactive in the synthesis of RNA. This suggestion has received strong support from electron microscope autoradiography, which has shown that RNA precursors are incorporated into the nucleus preferentially in the areas of the nucleoplasm (*1.33*) that are free of large chromatin aggregates. The means by which some parts of the genetic message come to be complexed as inactive chromatin aggregates is still obscure but is obviously fundamental to an understanding of cellular differentiation.

Autoradiographic studies have also shown that although RNA is synthesized chiefly within the nucleus, it can escape to the cytoplasm without any apparent rupture of the nuclear envelope (*1.34*). Some workers have suggested that the RNA molecules migrate from the nucleoplasm to the cytoplasm through the nuclear pores (Figure 6; surface view shown in Figure 13, inset). Unfortunately, it has not yet been shown conclusively that these structures are "pores" *in vivo.*

The nuclear envelope is a specialized part of the endoplasmic reticulum. Structures similar to nuclear pores are not seen in the rest of the endoplasmic reticulum, (Figure 15) and ribosomes (when bound to it) occur only on the cytoplasmic side of the nuclear envelope. When cisternae of rough endoplasmic reticulum are seen in surface view (Figure 14), many of the bound ribosomes are seen to lie in characteristic patterns (*1.35*). It is believed that these patterned arrays—the

13 **Part of a nucleus from a parenchyma cell of the oat coleoptile, showing the chromatin aggregates (Chr) and the two membranes of the nuclear envelope (NE). × 76,000. *Inset:* Nuclear pores seen in surface view in a mesophyll cell from an oat leaf. × 106,000. (Courtesy Dr. B. E. S. Gunning.)**

14 **Surface view of rough endoplasmic reticulum (ER) in a cell of the outer epidermis of the oat coleoptile, showing the arrangement of the ribosomes (R) bound to this membrane system. × 76,000. *Inset:* Detail of one of the polyribosomal aggregates shown in Figure 14. × 210,000.**

15 **The rough endoplasmic reticulum (ER) seen in profile in a cell of the outer epidermis of an oat coleoptile. Note that ribosomes (R) are absent from the vacuolar membrane (VM) which surrounds the vacuole (V). P, plastid. (Reproduced with permission from *Protoplasma*, 63, 1967, p. 405. × 76,000.**

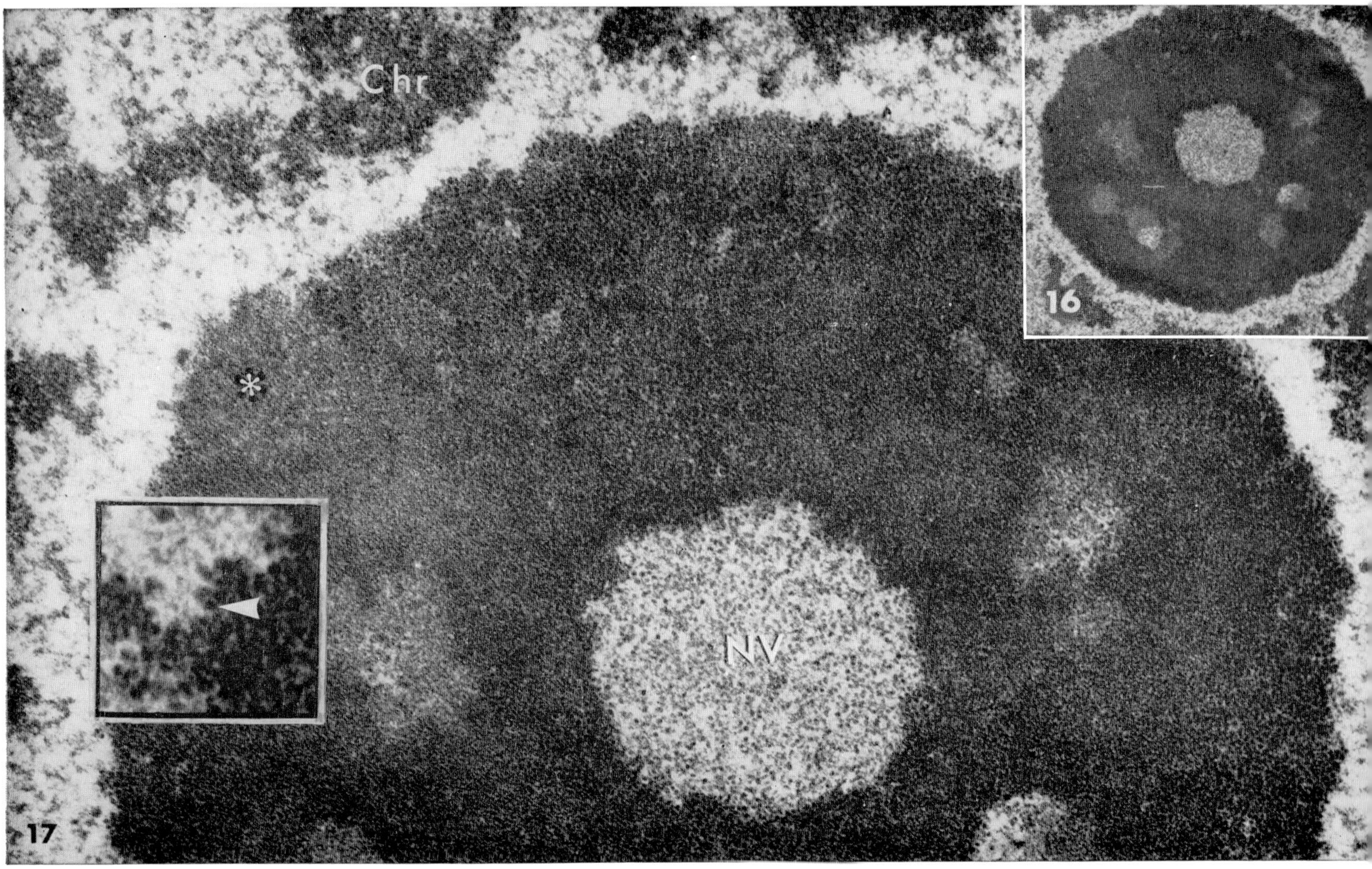

16 The nucleolus in a root tip cell of *Vicia faba.* Note the nucleolar vacuoles (compare with Figures 4 and 5). × 13,000. (Courtesy Dr. B. E. S. Gunning.)

17 Detail of Figure 16, showing the perinucleolar chromatin (Chr), nucleolar vacuoles (NV), and an unknown structure (asterisk). × 48,000. *Inset:* The proribosomes of the granular regions of the nucleolus. × 115,000. (Courtesy Dr. B. E. S. Gunning.)

polyribosomes—are ribosomes caught in the act of synthesizing protein (*1.36*) and aggregated one beside another along a strand of messenger RNA. Each ribosome has been shown to consist of two major subunits, and there is some evidence that the strand of messenger RNA passes from one ribosome to the next through the groove between the subunits. Even more remarkable is the possibility that proteins synthesized by these polyribosomal arrays upon the surface of the rough endoplasmic reticulum come to be sequestered *within* the cisternae of this membrane system.

Much of the evidence for these suggestions comes from studies of animal tissues, and there is pressing need for similar experiments to be carried out with plant cells, for the structures are common to both.

The Nucleolus

Despite moderately intensive study during the past ten years, the functions of the nucleolus (*1.37, 1.38, 1.39*) are still somewhat obscure. Indeed, the chemical composition of the various fibrous and granular components visible in the electron microscope (Figures 16 and 17) has not been established unequivocally. One important fact has emerged, however. The nucleolus seems to be a reservoir of RNA-rich particles that have been termed *proribosomes* (Figure 17 and inset). It is not yet known whether these particles are synthesized within the nucleolus or are merely stored there after synthesis within some other area of the nucleoplasm. These proribosomes are almost certainly discharged into the ground substance of the cytoplasm during mitosis, and it seems likely that they are converted there into ribosomes that can act in protein synthesis.

Plastids

Plastids occur in all cells of higher plants, although they are most conspicuous when they differentiate into *chloroplasts* in the photosynthetic parenchyma of leaves and stems (see Chapter 5). The photosynthetic pigments are absent, however, from the plastids of the tissues of young leaf primordia, shoot apices, roots, and the epidermis of most mature leaves and stems (see Figures 6, 15, and 35). Such plastids are often small and not readily distinguished from mitochondria in the living cell. They usually have only a sparsely developed system of internal membranes, though they often store large granules of starch. Starch-rich plastids are called *amyloplasts* and are abundant in the parenchyma cells of most plant organs. Although starch is also formed in chloroplasts as a result of photosynthesis, amyloplasts can make starch in total darkness, a fact that is essential for the storage of starch in underground organs.

General Comments

We have entered upon a reasonably detailed treatment of cell structure in this section with a very definite purpose in mind. In many of the later sections, morphological, physiological, and biochemical phenomena will be discussed in relation to organ or tissue structure; only rarely will it be possible to go into the fine structure of the cells concerned. By stressing at this stage the complexity of the biochemical machinery that carries out cellular physiology, it is our hope that you will remember that when a phenomenon has been "explained" at a high level of organization (organ or tissue), it is really the activity of this same complex cellular machinery that is involved.

GENERAL REFERENCES

BAKER, J. R. *Principles of Biological Microtechnique: A Study of Fixation and Dyeing.* London: Methuen & Co., Ltd., 1958.

BRACHET, J., and A. E. MIRSKY. *The Cell,* II. New York: Academic Press, 1961.

ESAU, K. *Plant Anatomy,* 2d ed., Chaps. 2 and 3. New York: John Wiley & Sons, Inc., 1965.

FAWCETT, D. W. *The Cell, an Atlas of Fine Structure.* Philadelphia: W. B. Saunders Co., 1966.

PORTER, K. R., and M. A. BONNEVILLE. *An Introduction to the Fine Structure of Cells and Tissues,* 2d ed. Philadelphia: Lea and Febiger, 1964.

ROELOFSEN, P. A. "The plant cell wall," in *Encylopaedia of Plant Anatomy,* III, Pt 4. Berlin-Nikolassee: Gebrüder-Borntraeger, 1959.

SAGER, R., and F. J. RYAN. *Cell Heredity.* New York: John Wiley & Sons, Inc., 1962.

SHARP, L. W. *Introduction to Cytology.* New York: McGraw-Hill Book Co., 1934.

WATSON, J. D. *Molecular Biology of the Gene.* New York: W. A. Benjamin, Inc., 1965.

WILSON, E. B. *The Cell in Development and Heredity,* 3d ed. New York: MacMillan Co., 1924.

REVIEWS AND RESEARCH PAPERS

1.1 SABATINI, D. D., K. BENSCH, AND R. J. BARNETT. "Cytochemistry and electron microscopy. The preservation of cellular ultrastructure and enzymic activity by aldehyde fixation." *J. Cell Biol.,* 17 (1963), 19–58.

1.2 PORTER, K. R., AND F. KALLMAN. "The properties and effects of osmium tetroxide as a tissue fixative with special reference to its use for electron microscopy." *Exp. Cell Res.*, 4 (1953), 127–141.
1.3 WATSON, M. L. "Staining of tissue sections for electron miscroscopy with heavy metals." *J. Biophys. Biochem. Cytol.*, 4 (1958), 475–479.
1.4 REYNOLDS, E. S. "The use of lead citrate at high pH as an electron opaque stain in electron microscopy." *J. Cell Biol.*, 17 (1963), 208–212.
1.5 PORTER, K. R. "The endoplasmic reticulum: some current interpretations of its forms and functions," in *Biological Structure and Function*, I, ed. T. W. Goodwin and O. Lindberg. New York: Academic Press, 1961.
1.6 BONNETT, H. T., AND E. H. NEWCOMB. "Coated vesicles and other cytoplasmic components of growing root hairs of radish." *Protoplasma*, 62 (1966), 59–75.
1.7 PORTER, K. R. "The ground substance: observations from electron microscopy," in *The Cell*, II, ed. J. Brachet and A. E. Mirsky. New York: Academic Press, 1961.
1.8 ROELOFSEN, P. A. "Ultrastructure of the wall in growing cells and its relation to the direction of growth," *Advances in Botanical Research*, 2 (1965), 69–145.
1.9 WARDROP, A. B., AND R. C. FOSTER. "A cytological study of the oat coleoptile," *Aust. J. Bot.*, 12 (1964), 135–141.
1.10 NEWCOMB, E. H. "Cytoplasm—cell wall relationships," *Ann. Rev. Plant Phys.*, 14 (1963), 43–64.
1.11 O'BRIEN, T. P., AND K. V. THIMANN. "Observations on the fine structure of the oat coleoptile. II. The parenchyma cells of the apex," *Protoplasma*, 63 (1967), 417–442.
1.12 WHALEY, W. G., H. H. MOLLENHAUER, AND J. H. LEECH. "The ultrastructure of the meristematic cell," *Amer. J. Bot.*, 47 (1960), 401–449.
1.13 HEPLER, P. K., AND E. H. NEWCOMB. "Fine structure of cell plate formation in the apical meristem of *Phaseolus* roots." *J. Ultrastr. Res.*, 19 (1967), 498–513.
1.14 LEDBETTER, M. C., AND K. R. PORTER. "A 'microtubule' in plant cell fine structure," *J. Cell Biol.*, 19 (1963), 239–250.
1.15 HEPLER, P. K., AND E. H. NEWCOMB. "Microtubules and fibrils in the cytoplasm of *Coleus* cells undergoing secondary wall deposition," *J. Cell Biol.*, 20 (1964), 529–533.
1.16 PICKETT-HEAPS, J. D., AND D. H. NORTHCOTE. "Relationship of cellular organelles to the formation and development of the plant cell wall," *J. Exp. Bot.*, 17 (1966), 20–26.
1.17 CRONSHAW, J., AND G. B. BOUCK. "The fine structure of differentiating xylem elements," *J. Cell Biol.*, 24 (1965), 415–431.
1.18 PICKETT-HEAPS, J. D., AND D. H. NORTHCOTE. "Cell division in the formation of the stomatal complex of the young leaves of wheat," *J. Cell Science*, 1 (1966), 121–128.
1.19 PICKETT-HEAPS, J. D. "The effects of colchicine on the ultrastructure of dividing plant cells, xylem wall differentiation and distribution of cytoplasmic microtubules," *Developmental Biology*, 15 (1967), 206–236.
1.20 "Mitochondrial structure and function," in *Biological Structure and Function*, II, ed. T. W. Goodwin and O. Lindberg. New York: Academic Press, 1961.
1.21 LEHNINGER, A. L., *The Mitochrondrion: Molecular Basis of Structure and Function*. New York: Benjamin Press, 1964.

1.22 PEACHEY, L. D. "Electron microscope observations on the accumulation of divalent cations in intramitochondrial granules," *J. Cell Biol.*, 20 (1964), 95–111.
1.23 PACKER, L. "Size and shape transformations correlated with oxidative phosphorylation in mitochondria," *J. Cell Biol.*, 18 (1963), 487–501.
1.24 LUCK, D. J. L. "Formation of mitochondria in *Neurospora crassa*," *Amer. Nat.*, 99 (1965), 241–253.
1.25 NASS, M. K., AND S. NASS. "Intramitrochondrial fibres with DNA characteristics," *J. Cell Biol.*, 19 (1963), 593–629.
1.26 SWIFT, H. "Nucleic acids of mitochondria and chloroplasts," *Amer. Nat.*, 99 (1965), 201–227.
1.27 MOLLENHAUER, H. H., AND D. J. MORRÉ. "Golgi apparatus and plant secretion," *Ann. Rev. Plant Phys.*, 17 (1966), 27–46.
1.28 MANTON, I., AND G. F. LEEDALE. "Observations on the fine structure of *Paraphysomonas vestita* with special reference to the Golgi apparatus and the origin of scales," *Phycologia*, 1 (1961), 37–57.
1.29 NORTHCOTE, D. H., AND J. D. PICKETT-HEAPS. "A function of the Golgi apparatus in polysaccharide synthesis and transport in the root-cap cells of wheat," *Biochem. J.*, 98 (1966), 159–173.
1.30 FRIEND, D. S., AND M. G. FARQUHAR. "Functions of coated vesicles during protein absorption in the rat *vas deferens*." *J. Cell Biol.*, 35 (1967), 357–376.
1.31 PRESCOTT, D. M. "The nuclear dependence of RNA synthesis in *Acanthameba* spp.," *Exp. Cell Res.*, 19 (1962), 29–34.
1.32 PRESCOTT, D. M., AND M. A. BENDER. "Synthesis of RNA and protein during mitosis in mammalian tissue culture cells," *Exp. Cell Res.*, 26 (1962), 260–268.
1.33 KARASAKI, S. "Electron microscopic examination of the sites of nuclear RNA synthesis during amphibian embryogenesis." *J. Cell Biol.*, 26 (1965), 937–958.
1.34 ZALOKAR, M. "Sites of protein and ribonucleic acid synthesis in the cell." *Exp. Cell Res.*, 19 (1960), 559–576.
1.35 BONNETT, H. T., AND E. H. NEWCOMB. "Polyribosomes and cisternal accumulations in root cells of radish," *J. Cell Biol.*, 27 (1965), 423–432.
1.36 SIEKEVITZ, P., AND G. E. PALADE. "A cytochemical study on the pancreas of the guinea pig. VI. Release of enzymes and ribonucleic acid from ribonucleoprotein particles," *J. Biophys. Biochem. Cytol.*, 7 (1960), 631–643.
1.37 HYDE, B. B., K. SANKARANARAYAN, AND M. L. BIRNSTIEL. "Observations on fine structure in pea nucleoli *in situ* and isolated," *J. Ultrastruct. Res.*, 12 (1965), 652–667.
1.38 LAFONTAINE, J. G. "The plant nucleolus," in *The Nucleus*, ed. A. J. Dalton and F. Haguenau. New York: Academic Press, 1968.
1.39 BIRNSTIEL, M. "The nucleolus in cell metabolism," *Ann. Rev. Plant Phys.*, 18 (1967), 25–58.

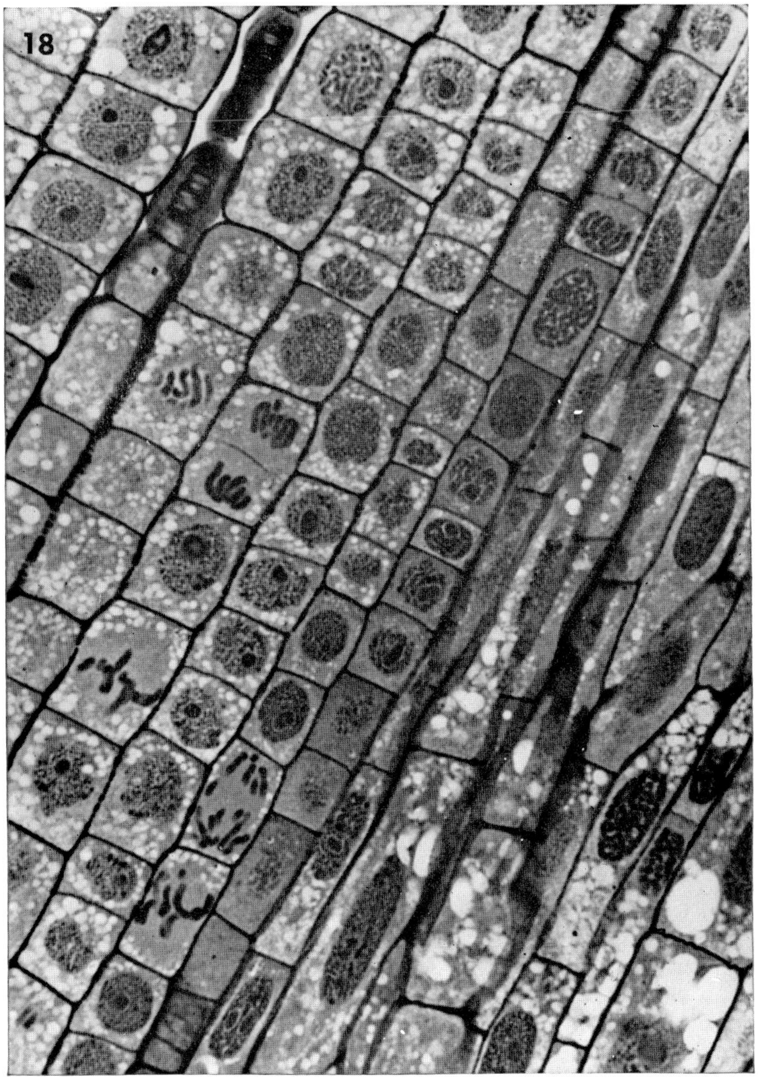

18 Longitudinal section through a root tip of onion showing cells in different stages of mitosis. The periodic acid/Schiff's reaction (PAS)/toluidine blue. × 300.

2

Cell Production: Mitosis

Meristematic cells in a rapidly growing root tip divide about once every 18–24 hours. Most of that time is spent in *interphase*, and it is upon interphase cells that we have concentrated in Chapter 1. The most dramatic cytological events, however, occur during *mitosis*, which lasts for about four hours in root-tip cells. Very detailed studies have been made of mitosis in living endosperm cells (*2.1, 2.2*). The description that follows and several of the illustrations are based largely upon these cells, although some stages are illustrated from root-tip cells (see Figures 18–34).

As the cell enters *prophase*, the first stage of mitosis, the chromatin, present as aggregates scattered throughout the nucleus in interphase, begins to condense and the coiled chromosomes make their appearance (Figure 19). At the poles of the nucleus, a clear zone develops (Figure 20, open arrows) and each chromosome can be seen to consist of a pair of threads, the *chromatids*, coiled about each other (Figure 21). Prophase ends, and *prometaphase* begins, often with dramatic swiftness, with rupture of the nuclear envelope and disappearance of the nucleolus (Figures 22 and 23).

It is at this time that *spindle fibers* first make their appearance. The use of interference contrast microscopy allows one to see these very thin fibers *in vivo* (Figure 24) (*2.3*). Just where they are synthesized is still obscure, though it is possible that they develop first in the clear zone during prophase (*2.4*). Each chromosome becomes attached to a spindle fiber at a special region, the *kinetochore* (or *centromere*; Figure 25). Not all spindle fibers appear to be attached to chromosomes; some extend right through the chromosome mass and are termed *continuous spindle fibers*.

During prometaphase, the chromosomes begin to migrate towards the *metaphase plate*, the plane at which separation of the chromatids will occur. This movement is far from regular; each chromosome may make several reversals of the direction of its movement before it finally comes to rest at the metaphase plate. As the chromosomes migrate, the arms tend to unravel and the double nature of their structure becomes more apparent (Figures 23 and 24). In thin sections each chromatid appears at this stage to be connected to its partner by thin bridges (Figure 23, arrows).

In the electron microscope, each spindle fiber is seen to contain a bundle of microtubules, morphologically similar to those that occur in the cell cortex during interphase (see Figures 6, 9, 25 and 27) (*2.5*). Most workers attribute the high refractive index and birefringence (*2.6*) of the spindle fibers to the microtubules. Just how these structures bring about chromosome movement in this and later stages is still one of the major mysteries of cell division.

At *metaphase*, the kinetochores of each chromosome usually lie in the metaphase plate. If the chromosomes are small (as in *Convolvulus*; see Figure 28), the whole chromosome set will lie essentially in a single plane, but if the chromosomes are large (as in *Haemanthus* and *Vicia*), only the kinetochore region will lie in the plane of the plate and the arms will project from the plate at right angles (Figure 26). In the electron microscope, each metaphase chromatid appears to be connected at the kinetochore to a spindle fiber, one from each pole of the cell (Figure 27). In time-lapse films of living cells, metaphase ends suddenly as the chromatids separate from one another and *anaphase* begins.

From this point on, each chromatid is called a *daughter chromosome*. These chromosomes migrate towards the poles of the spindle, guided apparently by their kinetochores, which "lead" during the movement (Figure 29). During late anaphase and early *telophase*, a new structure, the *phragmoplast*, begins to form between the two sets of daughter chromosomes. In sections it appears first as a row of densely stained nodules, with fibers extending from the nodules towards the poles of the spindle (Figure 30). When such a structure is examined in the electron microscope (Figure 31), the fibers are again seen to consist of microtubules and the nodules are regions of overlap between two sets of microtubules. The two sets appear to run in opposite directions and either arise or end within a mass of finely fibrous material that is probably responsible for the affinity of the nodules for stain in Figure 30.

In most cells, two processes—reformation of the daughter nuclei and formation of the cell plate within the phragmoplast—now commence. These two processes do not appear to be causally related. They simply overlap in time, for in many cells, formation of the cell plate is not complete for many hours after the daughter nuclei have regained their interphase appearance. During telophase, the chromosome mass disperses to give rise

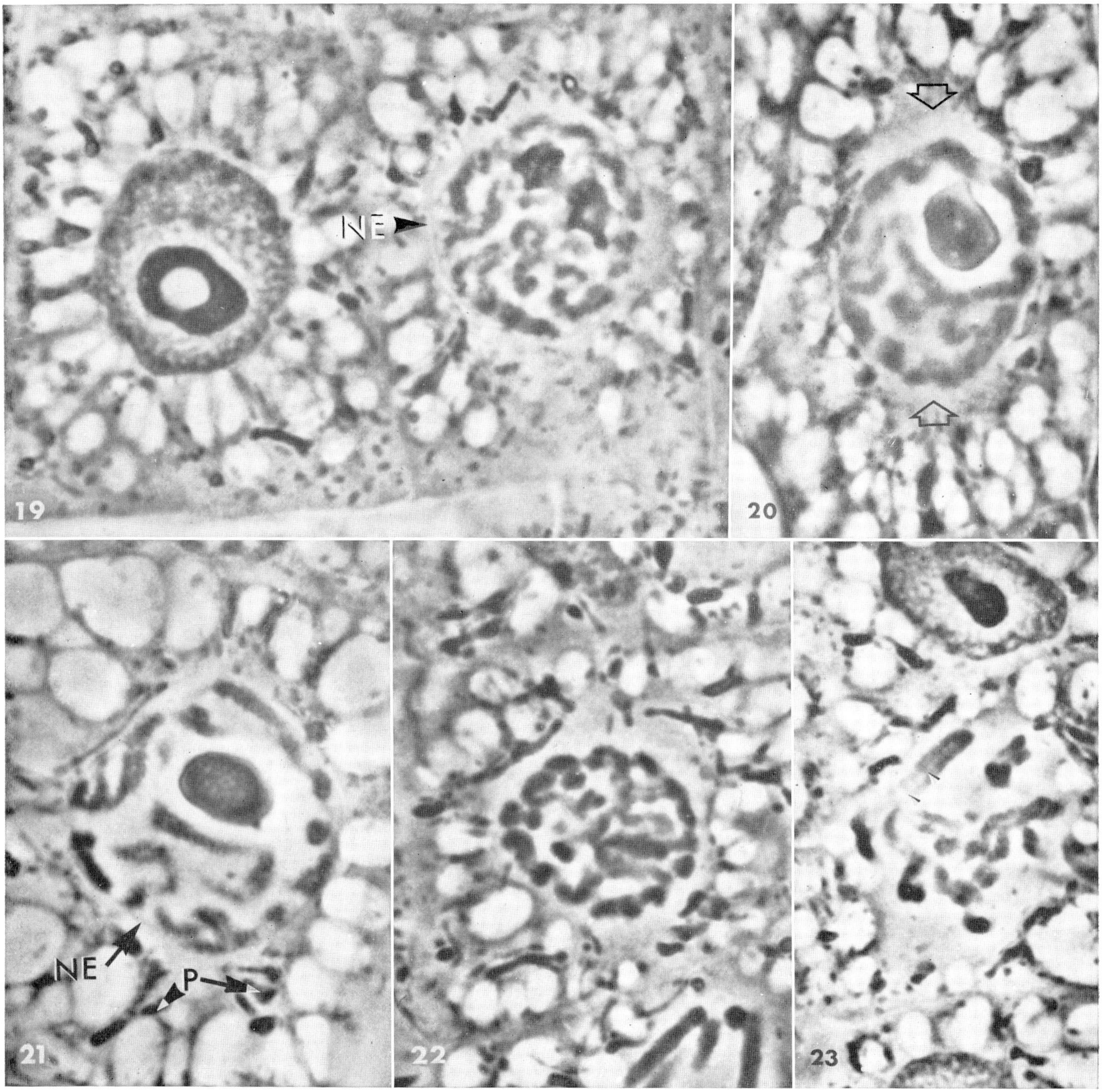

19-23 Phase contrast photomicrographs of nuclei from the root tip of *Vicia faba*, stained with iodine. All × 3150. (Specimen courtesy Dr. B. E. S. Gunning.)

19 Interphase (cell at left) and early prophase nucleus (cell at right). Note nuclear envelope (NE).

20 The "clear zones" (arrows) at the poles of an early prophase nucleus.

21 Late prophase, with chromatids visible, nuclear envelope (NE) intact. P, plastids.

22 Rupture of the nuclear envelope and disappearance of the nucleolus.

23 Early prometaphase; note the "bridges" (arrows) between the two chromatids of each chromosome.

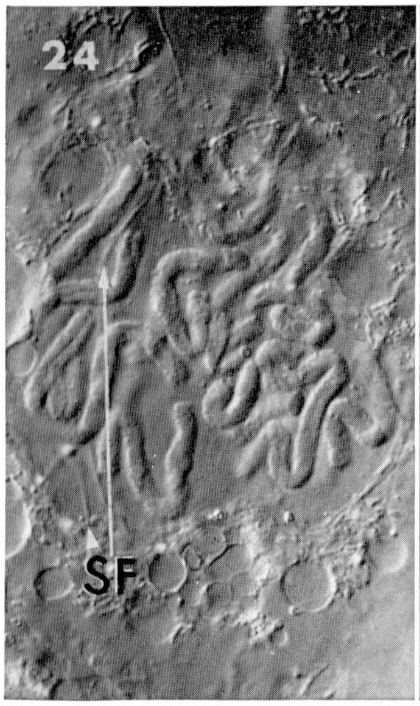

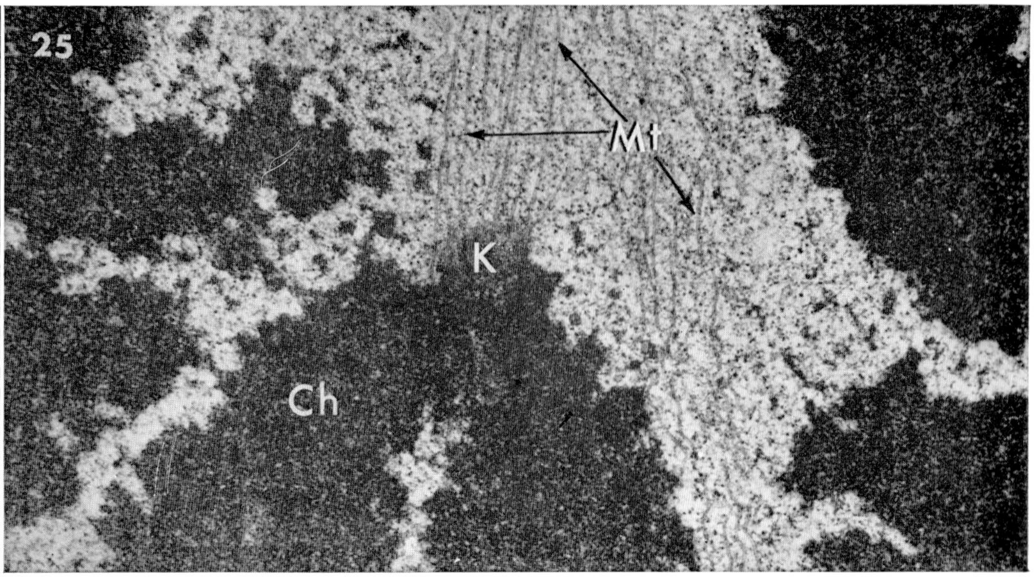

24 Interference contrast (Nomarski optics) photomicrograph of living cells of the endosperm of *Haemanthus* spp. (blood-lily) in prometaphase. Note spindle fibers (SF) and chromatids. × 450.

25 Electron micrograph of a stage comparable to that seen in Figure 24, showing chromosomes (Ch), kinetochore (K), and microtubules (Mt) of the spindle fibers. × 26,000.

26 Metaphase in living cells of *Haemanthus* endosperm. Interference contrast. SF, spindle fiber. × 450.

27 Electron micrograph of metaphase chromosomes (Ch). Each spindle fiber (SF) seen in the light microscope clearly contains a number of microtubules (Mt). Kinetochore (K). × 19,500. (Figures 24–27 all courtesy Dr. A. Bajer.)

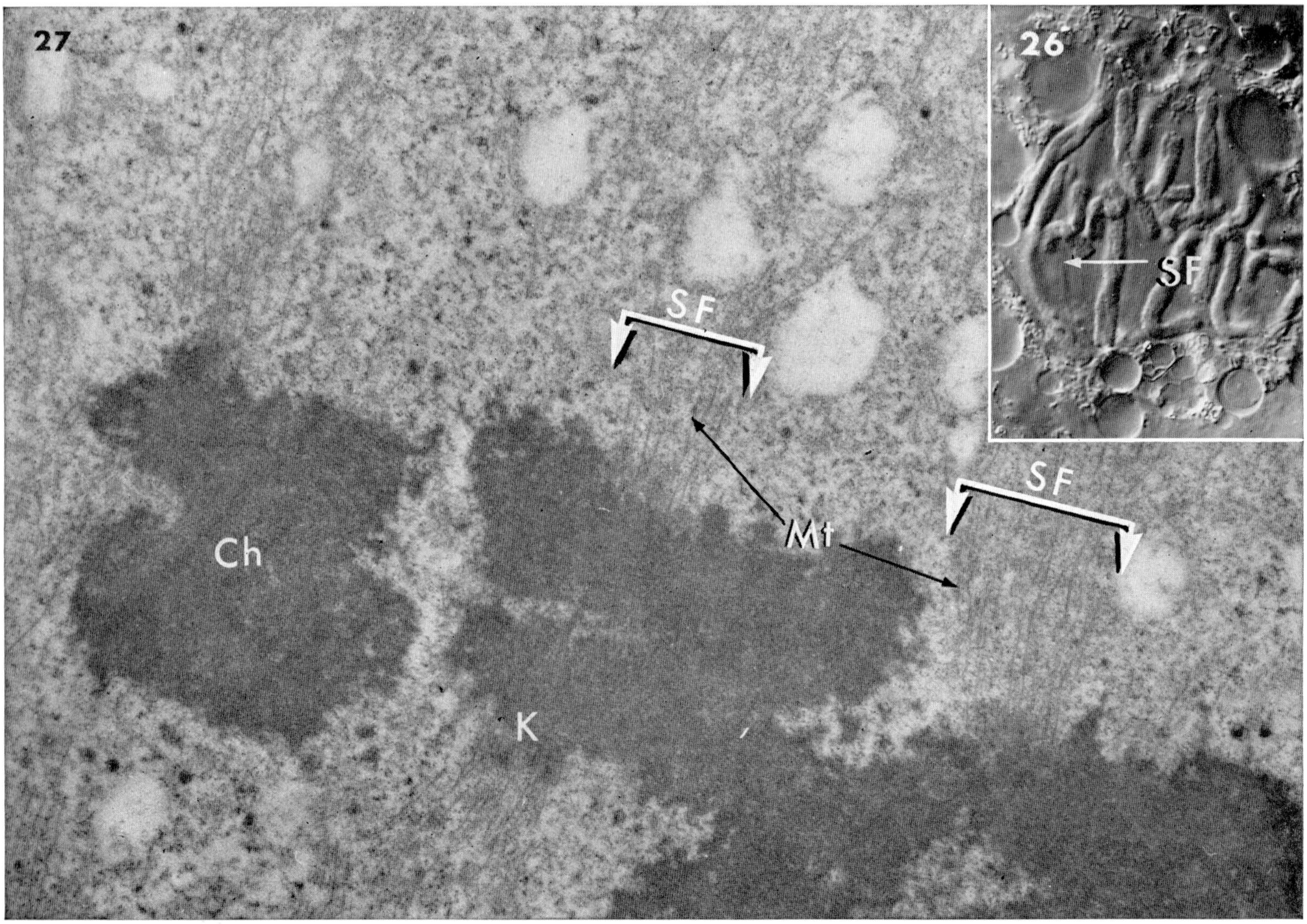

28 Early (upper cell) and mid-anaphase in cells from a *Convolvulus* spp. (bindweed) root tip. (Ch, chromosome.) Toluidine blue. × 1575.

29 Late anaphase in cells from root tip of *Vicia faba*. Note the kinetochore (K). Iodine-stained, phase contrast. × 4100. (Specimen courtesy Dr. B. E. S. Gunning.)

30 Late anaphase/early telophase in *Convolvulus* root-tip cells, stained with acid fuchsin and photographed in phase contrast. Note the central nodules to which spindle fibers (SF) of the phragmoplast are attached. × 3240.

31 An electron micrograph from *Haemanthus* endosperm at a stage judged to be closely similar to that shown in Figure 30. The central nodules of Figure 30 appear to be regions of overlap between the microtubules (Mt) of the phragmoplast. × 52,000.

32 Telophase nuclei and cell plate formation in cells from a *Convolvulus* root tip. CP, cell plate; N, nucleus; V, vacuole. Toluidine blue. × 1575.

33 and **34** Early (Figure 33) and late (Figure 34) stages in cell plate (CP) formation in living cells of *Haemanthus* endosperm. Interference contrast. N, nucleus. × 450. (Figures 31, 33 and 34 courtesy Dr. A. Bajer.)

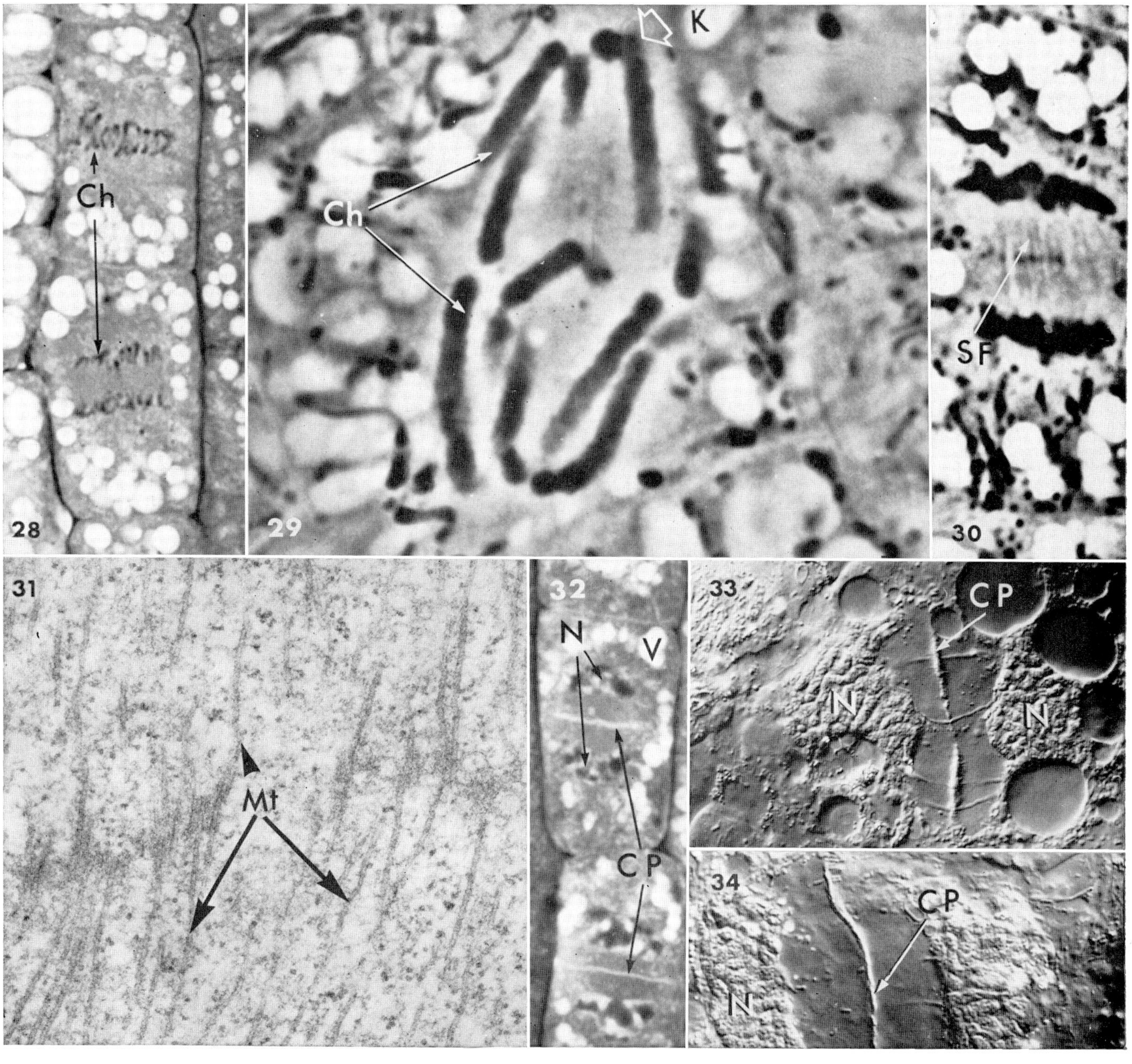

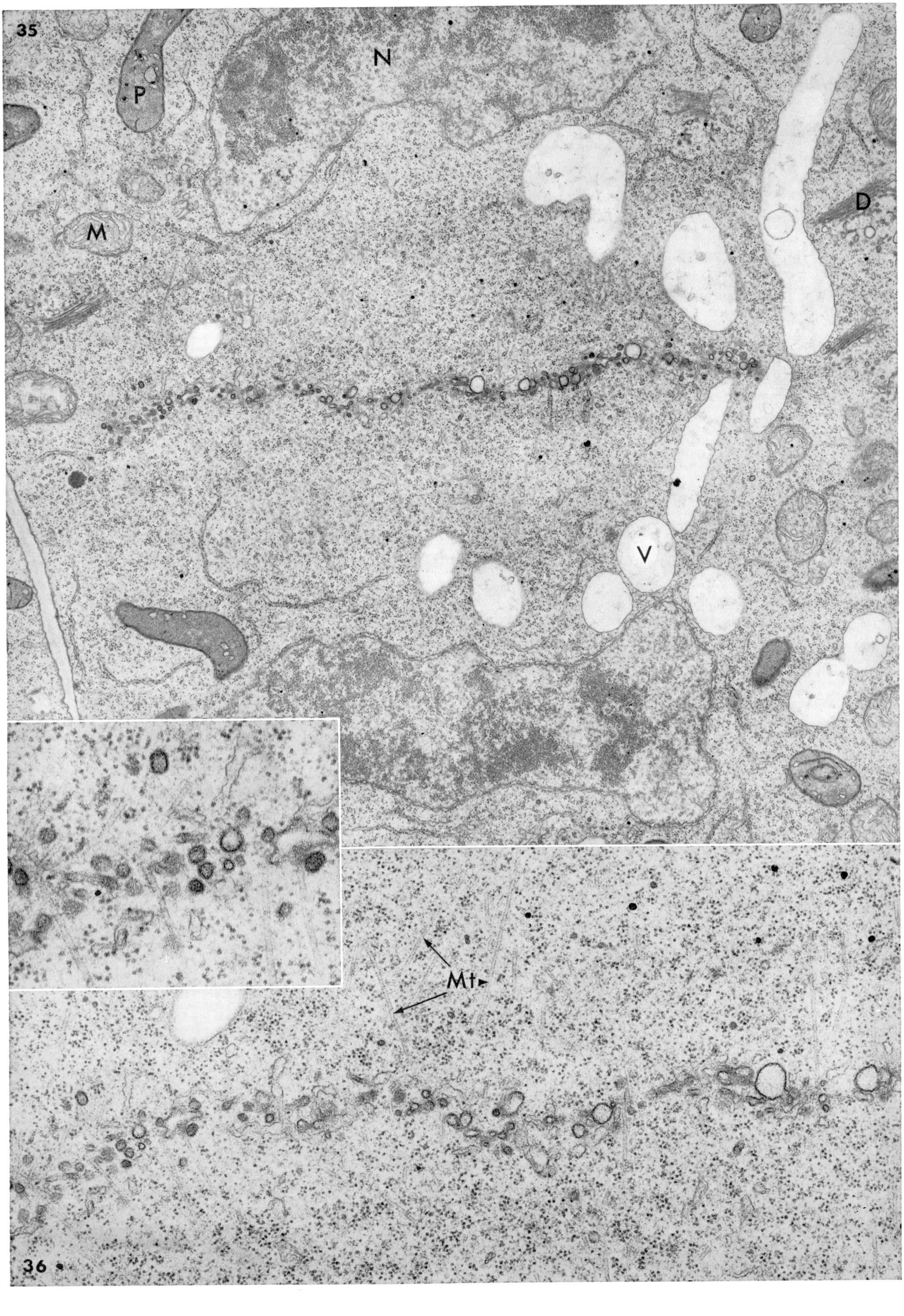
35
P
N
M
D
V
Mt
36

to chromatin, the nuclear envelope reappears, and nucleoli gradually re-form (Figure 32).

In root-tip cells, the cell plate expands radially from the center outwards, like the aperture of an opening iris diaphragm. When such a forming cell plate is examined in the electron microscope, numerous microtubules can be identified at the expanding margin of the plate (Figures 35 and 36). There is good evidence that the plate grows by the fusion of vesicles, perhaps derived in part from the dictyosomes, and it is believed that the microtubules of the phragmoplast somehow "sift" and "shuttle" these vesicles towards the growing perimeter of the plate (*1.13*). Indeed, when the forming cell plate is examined by interference contrast microscopy, structures similar to spindle fibers can be seen on either side of it (Figures 33 and 34). Many investigators have termed the vesicles that converge on the phragmoplast "pectin vesicles" for they believe that they are rich in methylated polyuronides and form the middle lamella of the future cell wall. These views are supported by the fact that quite mature cell plates fail to stain with cationic dyes (Figure 32; see also Figure 48). Since primary walls stain strongly with cationic dyes (Figure 5 and 28), either the methylated polyuronides are subsequently demethylated or polyuronides with free carboxyl groups (pectic acids) are added to the wall after the cell plate is completed.

Although it is likely that the cell plate is formed *in part* by fusion of dictyosome vesicles, such vesicles are not the only source of material from which the plate is constructed (*1.13, 2.7*). Figures 35 and 36 show that the forming plate is rich in cisternae of endoplasmic reticulum, and it is most likely that this membrane system is also playing an important role in cell plate formation.

Experiments with labelled precursors have shown that nuclear DNA synthesis is essentially complete by the onset of prophase. Thus, by the time chromosomes are apparent as distinct entities in the cell, duplication of the DNA for the daughter nuclei is largely finished (*2.8, 2.9, 2.10*). When the synthesis of RNA and protein was examined in a similar way, the astonishing fact emerged that these two major synthetic activities are strongly repressed during mitosis. It seems that the condensation of DNA into a chromosomal form prevents its activity as a template for RNA synthesis, the absence of which in turn limits the rate of protein synthesis. The period of greatest cytological activity—formation of chromosomes and their movements during prometaphase, metaphase, and anaphase—occurs when the cell's synthetic activities are at a minimum. It is the interphase cell (ironically called the "resting" cell by the early microscopists) in which metabolism is most active, whereas a cell in mitosis can reasonably be termed "physiologically enucleated" (*1.32*).

GENERAL REFERENCES

Brachet, J., and A. E. Mirsky. *The Cell*, III. New York: Academic Press, 1961.
Hughes, A. *The Mitotic Cycle.* London: Butterworth's Scientific Publications, 1952.

35 and **36** **Electron micrograph of cell plate formation in the root-tip cells of *Phaseolus*. This cell is judged to be at a stage comparable to that shown in Figure 32, but the section has passed through the edge of the phragmoplast, not through the center. The plate consists chiefly of vesicles of irregular shape, limited by a membrane of high electron contrast. The microtubules (Mt) of the phragmoplast are interspersed among these vesicles, as are numerous profiles of endoplasmic reticulum (ER). At the extreme margin of the plate (inset) discrete vesicles (believed to be dictyosome vesicles) appear to be fusing to form the large, more irregular vesicles that are characteristic of the central region of the plate. It is believed that the microtubules direct these dictyosome vesicles to the margin of the growing cell plate. M, mitochondria; N, nucleus; V, vacuole; P, plastid. Figure 35, × 22,500; Figure 36, × 49,500; inset, × 86,700. (Courtesy Dr. E. H. Newcomb.)**

LOCKE, M. *Reproduction: Molecular, Subcellular, and Cellular.* 24th Symposium of the Society for Developmental Biology. New York: Academic Press, 1965.

"MITOSIS," in *The Journal of Cell Biology, 25* (1965), 1–167.

SHARP, L. W. *Introduction to Cytology.* New York: McGraw-Hill Book Co., 1934.

WHITE, M. J. *The Chromosomes,* 5th ed. London: Methuen & Co., Ltd., 1961.

WILSON, G. B., AND J. H. MORRISON. *Cytology,* Chaps. 5 and 6. London: Reinhold Publishing Corp., 1961.

REVIEWS AND RESEARCH PAPERS

2.1 BAJER, A. "Cine micrographic analysis of cell plate formation in endosperm," *Exp. Cell Res.,* 37 (1965), 376–398.

2.2 BAJER, A., AND J. MOLÉ-BAJER. "Cinematographic studies on mitosis in endosperm. II. Chromosome, cytoplasmic and Brownian movements," *Chromosoma* (Berl.), 7 (1956), 558–607.

2.3 BAJER, A., AND R. D. ALLEN. "Role of phragmoplast filaments in cell-plate formation," *J. Cell Science,* 1 (1966), 455–462.

2.4 DUNCAN, R. E., AND M. D. PERSIDSKY. "The achromatic figure during mitosis in maize endosperm," *Amer. J. Bot.,* 45 (1958), 719–729.

2.5 LEDBETTER, M. C., AND K. R. PORTER. "A 'microtubule' in plant cell fine structure," *J. Cell Biol.,* 19 (1963), 239–250.

2.6 INOUÉ, S. "Organization and function of the mitotic spindle," in *Primitive Motile Systems in Cell Biology,* ed. R. Allen and N. Kamiya. New York: Academic Press, 1964.

2.7 PORTER, K. R., AND R. D. MACHADO. "Studies on the endoplasmic reticulum. IV. Its form and distribution during mitosis in cells of the onion root tip," *J. Biophys. Biochem. Cytol.,* 7 (1960), 167–180.

2.8 HOWARD, A. "The physiology of mitosis," *Encylopaedia of Plant Physiology,* XIV, ed. W. Ruhland, Berlin: Springer-Verlag, 1961.

2.9 TAYLOR, J. H. "The mode of chromosome duplication in *Crepis capillaris,*" *Exp. Cell Res.,* 15 (1958), 350–357.

2.10 STERN, H. "The regulation of cell division," *Ann. Rev. Plant Phys.,* 17 (1966), 345–378.

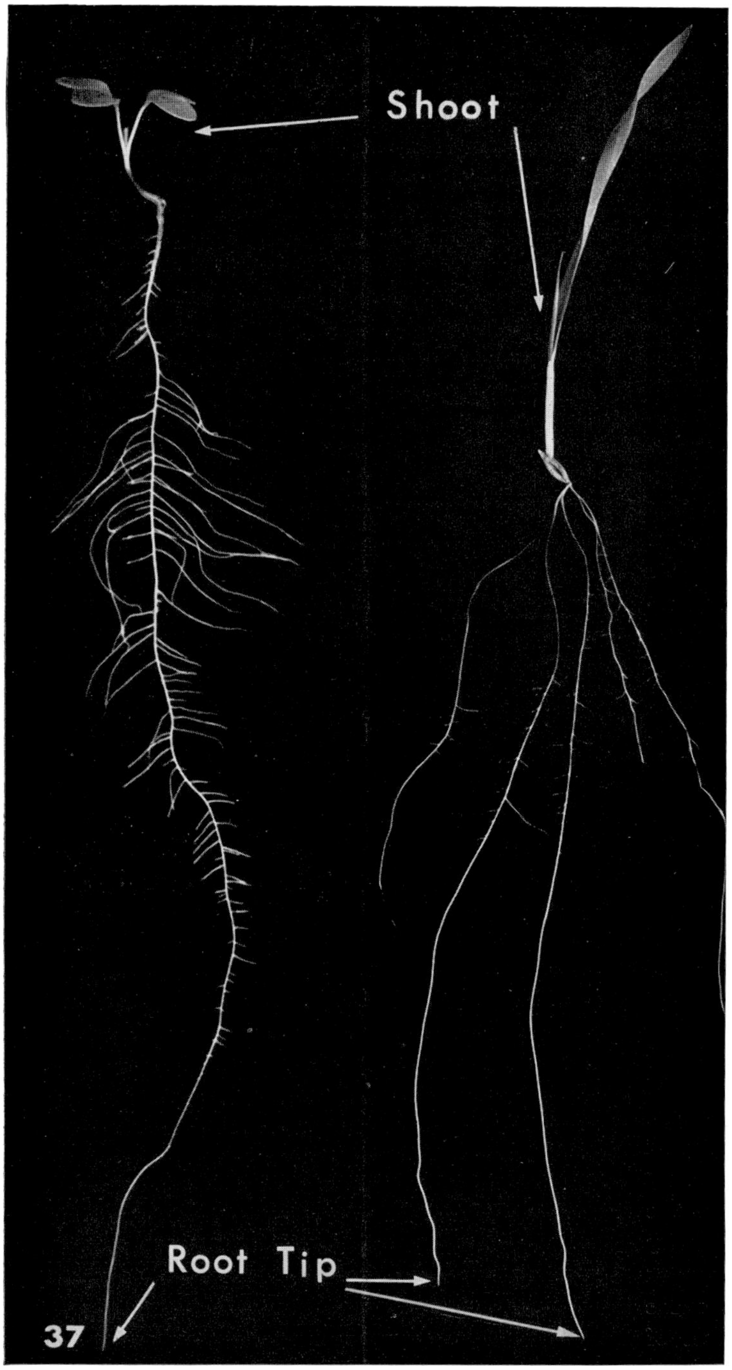

37 Seedlings of *Raphanus sativus* (radish) (left) and *Hordeum vulgare* (barley) (right). Contrast the root system of radish (tap-root and lateral roots) with the five seminal roots and laterals of the barley. × 0.7.

3

The Root

All green plants are photosynthetic autotrophs; that is, they can synthesize all of the complex organic molecules of which they are composed when supplied only with light, carbon dioxide, and an aqueous solution of appropriate inorganic ions. In terrestrial higher plants, the plant body is differentiated into a *root system* and a *shoot system*, neither of which, considered by itself, is autotrophic. In the intact plant, each of these two systems depends upon the proper functioning of the other for its own growth and development, and each regulates, and is in turn regulated by, the activities of the other *(3.1)*. This point is stressed because in treating each system and its parts separately, it is easy to overlook the mutual interdependence of the parts.

Absorption of water and solutes, anchorage and support of the plant are major functions of the root system. Since terrestrial higher plants colonize a wide variety of soils in different climates, it is not surprising that the gross morphology of the root system varies among different species and even among members of the same species grown in different habitats (compare the seedling root systems of radish and barley in Figure 37). Yet, in spite of these variations, the individual roots of any root system have many important features in common.

Simple marking experiments show that all roots grow at their tips. In most roots, the extreme apex is covered by a conical mass of cells, the *root cap*, which protects the growing point as it is thrust through the soil. Many roots also develop *root hairs* (Figure 38), tubular outgrowths from the cells of the *root epidermis*. Although root hairs increase the absorptive area of the root, early work suggests that the root-hair surface is not more permeable to aqueous solutions than the rest of the surface of the same cells *(3.2)*.

Several complex events—cell division, cell growth, and cell differentiation—occur in the root tip. Unfortunately, some texts treat the root tip as if these events occurred in sequence. In fact, as an examination of Figure 39 will confirm, they occur in parallel, albeit at different rates at different levels in the tip. Thus, *mitotic figures* are present in the highly vacuolated cells of the *cortex* in Figure 39D well above the level at which the first *sieve tubes* have been fully differentiated.

If in Figure 39 one traces the files of cells that form the differentiating cortex and *stele* downwards towards the root cap, the files are seen to converge upon a group of cells in the midline of the root. For many years, these cells were thought to be the *meristematic initials* from which all of the cells of the epidermis, cortex, and stele were derived. It is now known that these cells lie within the *quiescent center*, a region in which DNA duplication and mitosis occur at very low rates (Figures 40 and 41). The true meristematic initials lie at the surface of the roughly hemispherical group of cells that forms the quiescent center. It is these meristematic cells, rather than those of the quiescent center, that by growth and division produce the cap, epidermis, cortex, and stele. There is evidence that the cells of the quiescent center do divide to replace the meristematic cells if the latter are damaged *(3.3, 3.4)*.

Cell growth appears to accompany cell division at all levels in the root tip for there is no zone in which cell size decreases. The divisions that lead to the formation of new files of cells parallel to the long axis of the root appear to cease quite close to the margin of the quiescent center. Thus, the bulk of the increase in diameter of the root, brought about by radial enlargement of cells, is unaccompanied by cell division.

Cell differentiation clearly begins close to the meristematic initials, for the pattern of the future stele is quite evident as a difference in the intensity of staining in Figure 39C, a transverse section which corresponds to level C in Figure 39D. Note also that gas spaces are quite well-developed at that level. Divisions at right angles to the axis of the root cease earlier within the stele than within the cortex, so that the *provascular tissue* comes to be composed of rather elongated cells. Sieve tubes are the first fully differentiated cell type to appear in the root and make their appearance at level B.

Between levels B and A the rate of cell extension rises sharply and the cells become more and more vacuolated. The cortical cells that lie close to the stele vacuolate most rapidly (the vacuolated cells that lie at the surface of the root in Figure 39 are remnants of a persistent root cap; the epidermis is the first layer of densely stained cells beneath this persistent root cap).

Mature Tissues of the Primary Root

Beyond a certain distance from the apex, further extension of the cells ceases. The distance from the

38 A root tip of a radish seedling, showing the root cap and root hairs at various stages of development. × 15.

tip at which this point is reached varies with species and the conditions of growth. In rapidly growing roots the zone of elongation may be more than 20 mm long. Above the zone of elongation, all roots show a characteristic pattern in transverse section (Figure 42). They are clearly differentiated into a cortex of highly vacuolated parenchyma cells and a stele that is separated from the cortex by a special layer of cells, the *endodermis.* The cell walls of many of the endodermal cells are impregnated with lipid. This lipid may be present as a suberized strip within the radial walls (termed the *Casparian strip*) or the whole of the wall may be impregnated, as in the pea root (Figure 45). The presence of the lipid prevents the normal reaction of the polyuronides of the wall with

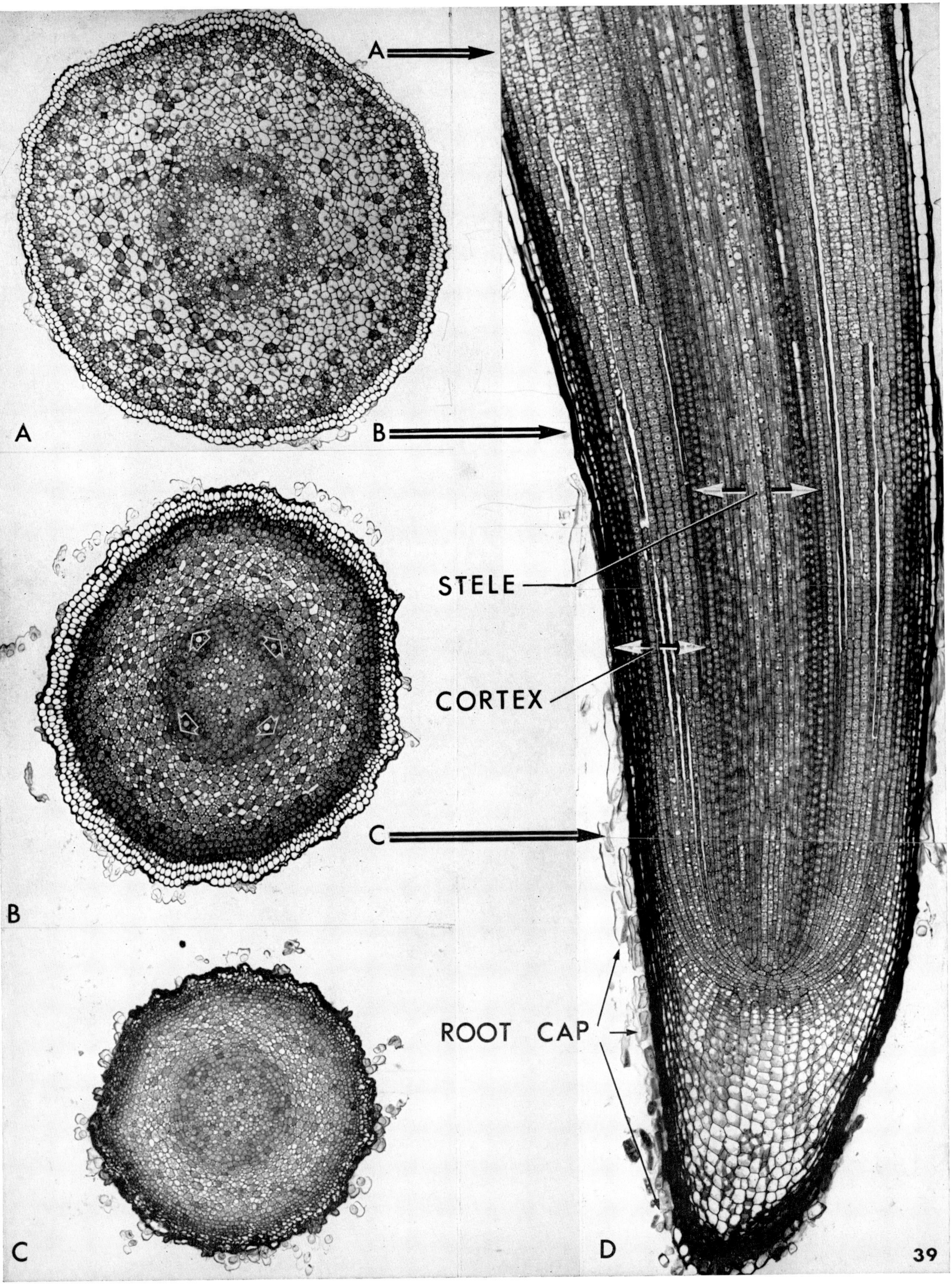
A
B
C
D
STELE
CORTEX
ROOT CAP

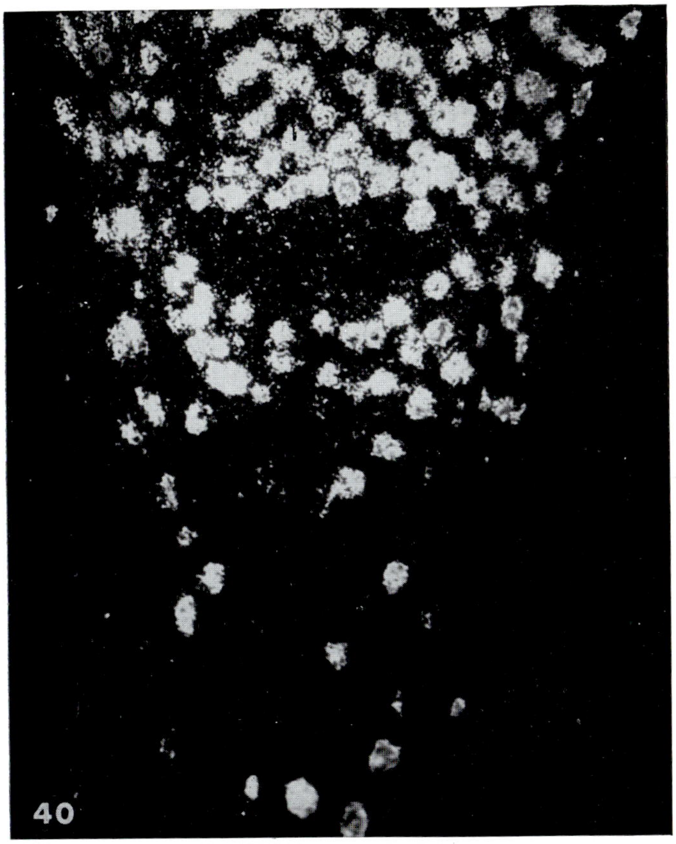

40

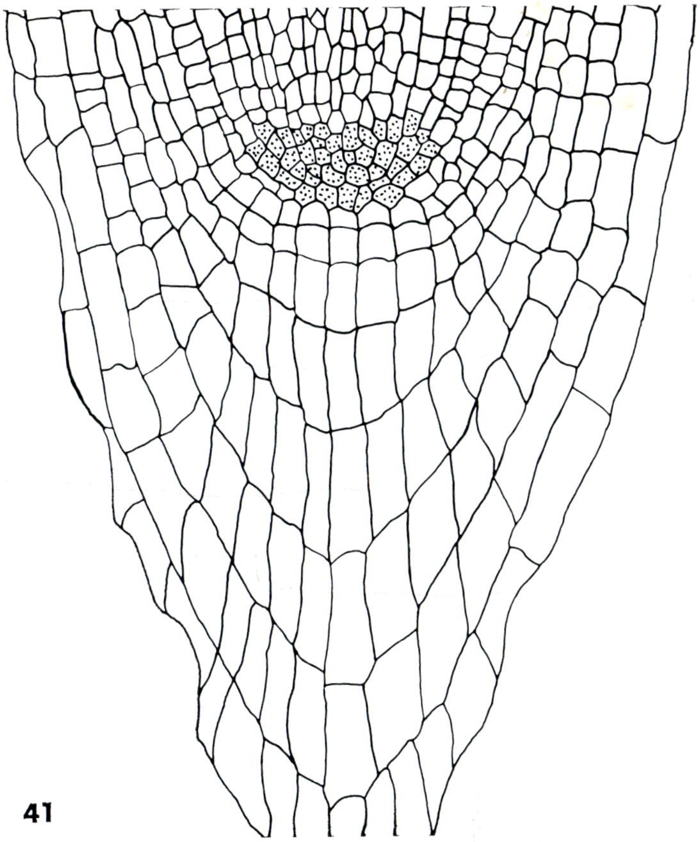

41

40 and 41 The quiescent center of the root apex of *Sinapis alba* (white mustard), seen as an autoradiogram (Figure 40), and as a diagram (Figure 41). Both figures show the distribution of the cells in the root apex which fail to incorporate ^{3}H-thymidine, indicating low rates of DNA synthesis in these cells. Figure 40 is photographed by dark field illumination so that the silver grains appear white on a black background. Figure 40, × 550; Figure 41, × 360. (Both courtesy Dr. F. A. L. Clowes.)

cationic dyes (see Figure 45, arrows). In many monocotyledons, the endodermal cells have asymmetrically thickened walls that may be lignified and suberized; they present a spectacular sight in polarized light (see Frontispiece). It is usually stated that this lipidic impregnation of the wall controls the passage of aqueous solutions from the cortex to the stele, but definite evidence to support this view is scanty. The stele is bounded by the *pericycle*, a layer of vacuolated parenchyma cells that retain their meristematic potential and are involved in the early stages of lateral root initiation (see Figures 48 and 49). In roots that have secondary growth, both the vascular cambium and the cork cambium can arise from the pericycle.

The xylem and the phloem are arranged upon alternating radii within the stele. In dicotyledons the xylem elements extend along the radii into the center of the root (Figure 42). In a few monocolytedons a distinct xylem strand forms in the

39 Longitudinal section (D) through the root-tip of a cultured *Convolvulus* root. A, B, and C illustrate the appearance of a comparable root in transverse section, taken at levels that correspond to the labelled arrows of the longitudinal section. The pattern of the future stele is evident in transverse section quite close to the root apex, as are the gas spaces of the cortex. The first provascular cells to mature in the stele are the four sieve tubes (within the open arrows in B). Vacuolation of the cortex proceeds centrifugally (from stele towards epidermis) as one proceeds from the root apex towards the base (compare C with B and A). The large, vacuolated cells at the surface of the root in all four photographs are part of the root cap which persists over the epidermis of the root under these conditions of culture. The periodic acid/Schiff's reaction (PAS)/toluidine blue. × 140.

center of the stele, but in other members of the group this area is filled with parenchyma. Figures 43, 44, and 46 illustrate the cell types that make up the xylem and phloem poles in a pea root. In each case, the earliest cells to differentiate (the *protoxylem* and *protophloem*) lie towards the pericycle, whereas the later formed cells (the *metaxylem* and *metaphloem*) lie towards the center of the root. Note that the width of the pericycle is not constant but varies from a single-celled layer over the xylem poles to two to three cells thick at the phloem poles.

The xylem contains *tracheary elements* (through which water and salts are moved) and numerous *xylem parenchyma* cells. The phloem contains sieve tubes, *phloem parenchyma* cells and *fibers*. These fibers help to support and stiffen the root as it penetrates into the soil. Unfortunately, little is known of the functions of the xylem and phloem parenchyma cells. In view of their position in the root, it is likely that they help to transfer nutrients from the vascular system to the growing cells of the meristem, and to pump solutes into the transpiration stream for transfer to the shoot. The structure and function of tracheary elements and sieve tubes is treated in more detail in Chapter 7.

The pattern of tissue differentiation in the primary root is remarkably precise and has stimulated a good deal of interest. How are the products of the meristem constrained to follow such precise pathways of cellular differentiation?

Although root systems engage in a complex nutritional interplay with the shoot in the intact plant, it has proved possible to culture isolated roots of certain species in a mineral salts medium enriched with sugar and appropriate vitamins. Some of these cultures have been maintained continuously for more than thirty years (*3.5, 3.6*). Isolated root cultures of this type have allowed experiments on the factors that control the development of the mature tissue patterns of the primary root (*3.7*). Even when very small (0.5 mm) tips were cultured, the same pattern of mature tissues was produced. Clearly, this suggests that control of the mature tissue-pattern resides within the group of cells that includes the quiescent center and the highly meristematic cells at its periphery. The pattern does not seem to be im-

42-46 **The primary vascular tissues of a root of *Pisum sativum* (pea).**

42 **Transverse section of a primary root of pea. PAS/toluidine blue. × 68.**

43 **A single "pole" of xylem from the root shown in Figure 42. The narrow protoxylem tracheary elements (small asterisks) lie nearer to the pericycle (Pe) and are readily distinguished from the wider, more recently differentiated metaxylem elements (large asterisks). Note the abundance of xylem parenchyma cells that ensheath the tracheary elements. The center of the root lies towards the bottom of the photograph. Toluidine blue. × 1680.**

44 **Part of the primary phloem in the root shown in Figure 42. It is difficult to decide just which cells are sieve tubes for the pericycle has proliferated and young fibers can be seen in the process of differentiating their thick secondary walls (DPhF). En, endodermis. The center of the root is towards the bottom of the photograph. Toluidine blue. × 1680.**

45 **Part of the endodermis in the root shown in Figure 42. These endodermal cell walls do not have Casparian strips; rather, the whole wall is impregnated with lipid. Such lipid-rich walls fail to stain with toluidine blue (small arrows) and at first sight the cytoplasm appears to be plasmolysed (see also Figure 56). The irregular outline of the endodermal cell walls is the result of crushing due to the expansion of the stele. The center of the root lies towards the *left* side of the photograph. Toluidine blue. × 1680.**

46 **Transverse section through part of the mature tissues of a pea root, showing an early stage in the development of the vascular cambium (CZ), and the arrangement of the mature tissues of the primary xylem (PXP) and phloem (Ph). A sieve plate in a metaphloem element is shown in higher magnification in the inset. Endodermis, En; Pericycle, Pe. MXV, metaxylem vessel; DMXV, differentiating metaxylem vessel. Toluidine blue. × 825; inset, × 2000.**

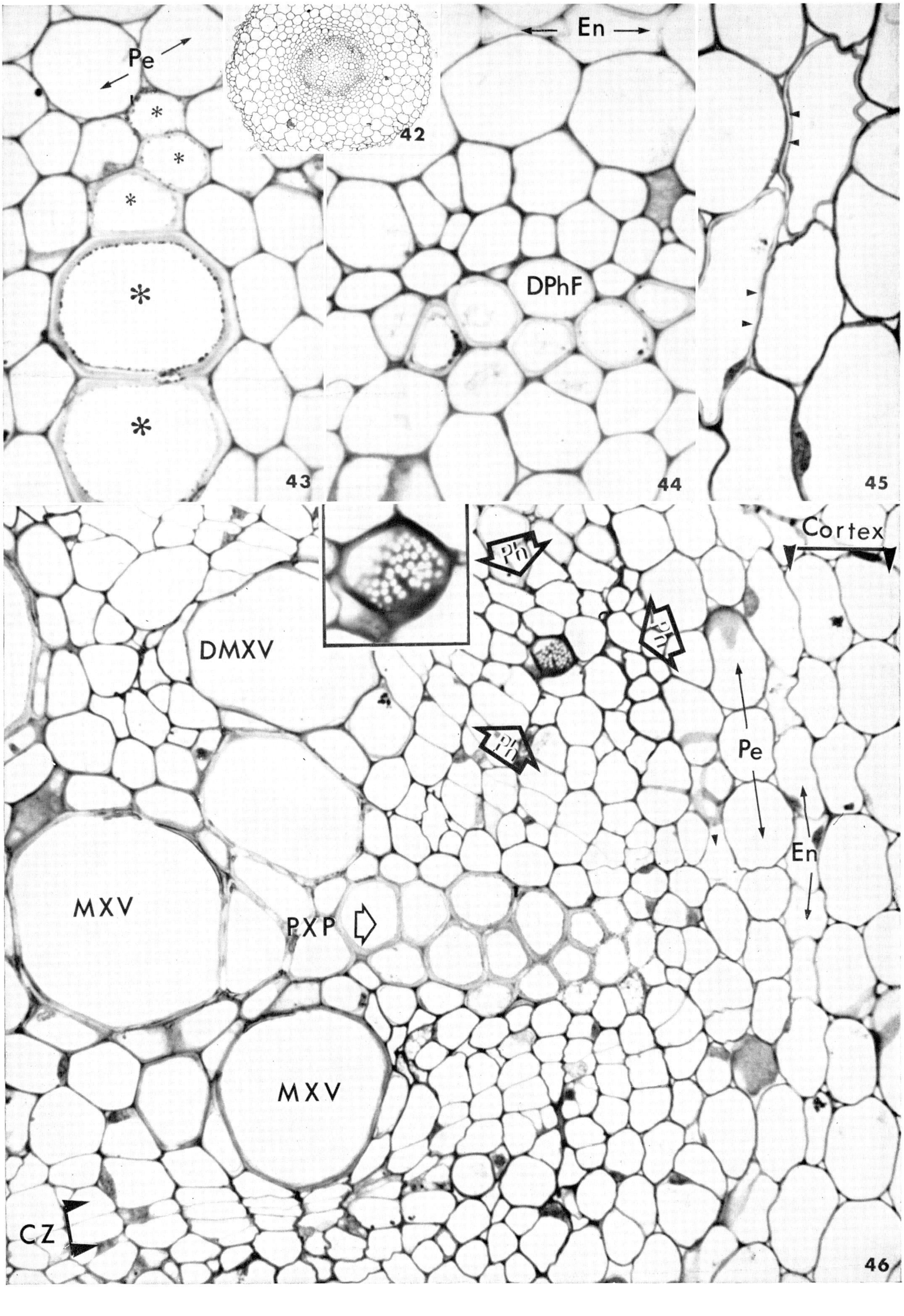
Pe
*
*
*
*
*
42
43
En
DPhF
44
45
Cortex
Ph
Ph
Ph
DMXV
Pe
En
MXV
PXP
MXV
CZ
46

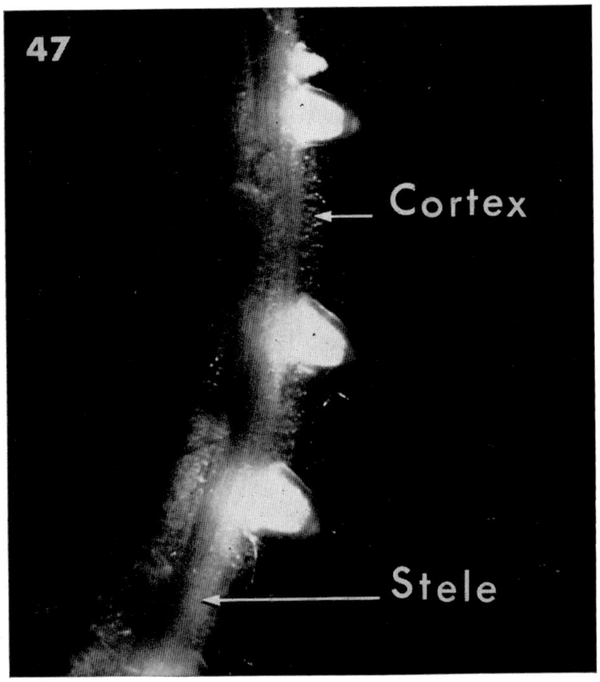

47 **Lateral root primordia in a freshly dissected root of *Zea mays* (maize). Most of the cortex has been peeled away to expose the stele. × 30.**

posed directly by influences (nutritional or hormonal) that come from the fully differentiated cell types, for the pattern is maintained in their absence. Nor can the pattern be related in any simple way to the shape of the quiescent center and its associated meristematic cells. A longitudinal section of a maize root tip appears remarkably similar to that of the *Convovulus* root shown in Figure 39D, yet the vascular patterns produced are quite different. Only one clue to the control mechanism has emerged in these studies. It has been shown that the *number* of arcs of xylem and phloem produced by a cultured root tip is sensitive to the concentration of auxin (indole acetic acid) in the medium. This is just one of the many facets of auxin action and presumably many biochemical steps intervene between the initial action of the hormone and the morphological expression of that action.

Within the stele, production of new cells and differentiation of new vascular elements continue well beyond the region of maximum extension. These elements that differentiate under conditions of lowered growth rate, the metaxylem and metaphloem, are often of wider diameter than those of the protoxylem and protophloem (Figure 46). In addition, a *cambium*—a new and rather different type of meristem—forms in many roots during or after the formation and maturation of the metaxylem and metaphloem and eventually gives rise to *secondary xylem* and *secondary phloem* (see also Figures 53, 54, and 55).

48-50 **Stages in lateral root initiation in maize.**

48 **The first divisions of the pericycle, which mark the initiaton of a lateral root primordium. The sieve tube (ST) marks the position of the stele. Note the unstained cell plate (white arrows). Longitudinal section, toluidine blue. × 1260.**

49 **A later stage in the development of a root primordium which shows in radial longitudinal section the "nest" of dividing cells, both in the stele and in the cortex. The pericycle (Pe) lies between the proliferating cells of the cortex on the right, while a number of sieve tubes (ST) lie between the pericycle and the proliferating cells of the stele on the left. Toluidine blue. × 690.**

50 **Longitudinal section of a lateral root, just at the point of its emergence from the cortex of the mother root which has been sectioned transversely (compare with Figure 47). A layer of remarkable secretory cells lies over the surface of the primordium, and the cells of the cortex are collapsed wherever they are in contact with this layer (asterisks). The inset shows the thick layer of extracellular material (ECM) over the surface of these cells and also some secreted product in the apical pole of the cells. The presence of a phragmoplast (inset, white arrows) in these secretory cells indicates that they grow and divide to keep pace with the expansion of the surface of the developing root primordium. En, endodermis; Pe, pericycle; ST, sieve tube; MXV, metaxylem vessel. PAS. × 375; inset × 1900. (Specimens courtesy Mr. F. T. Bellware and Mr. G. Glantz.)**

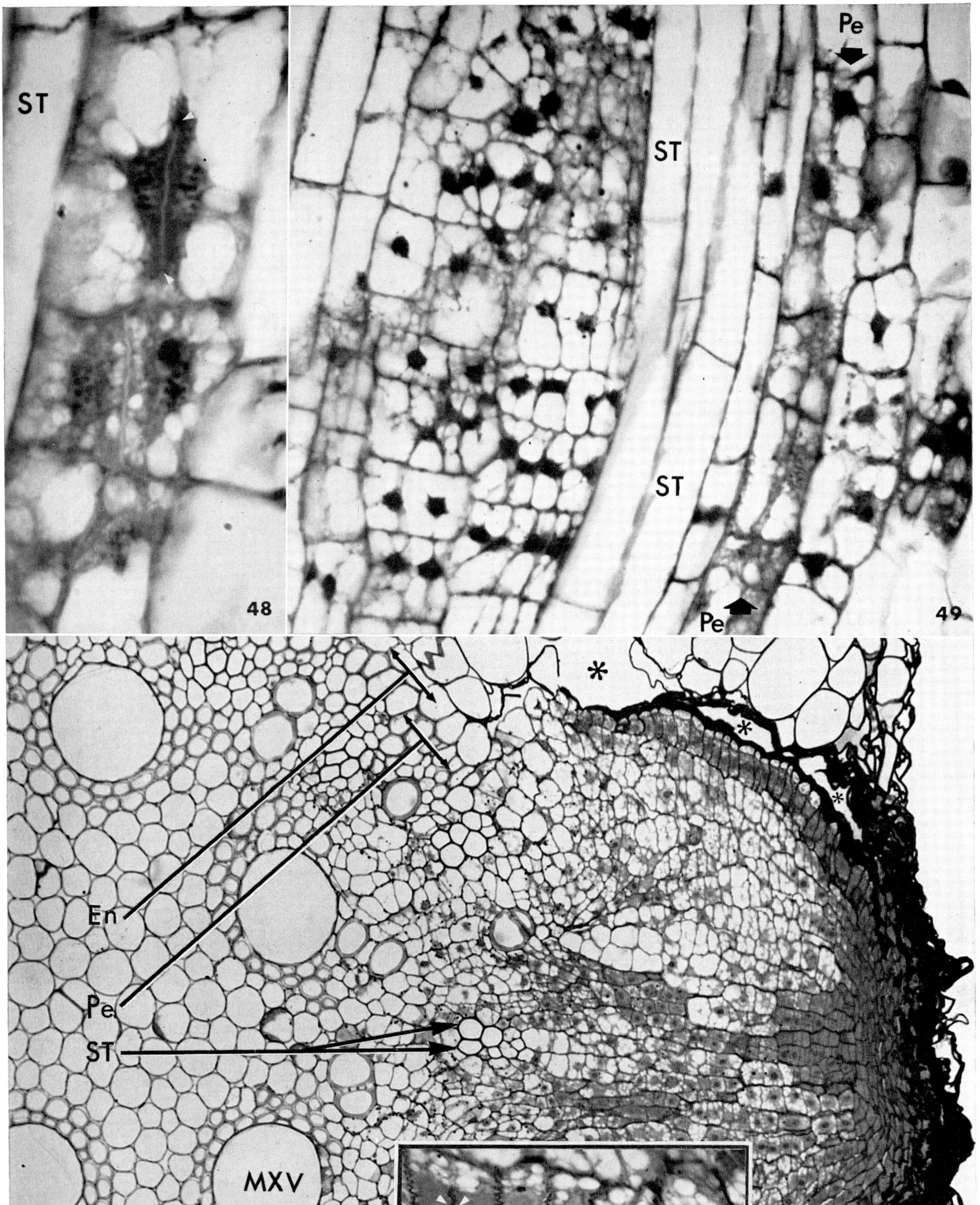
ST
48
ST
ST
Pe
Pe
49
En
Pe
ST
MXV
ECM
50

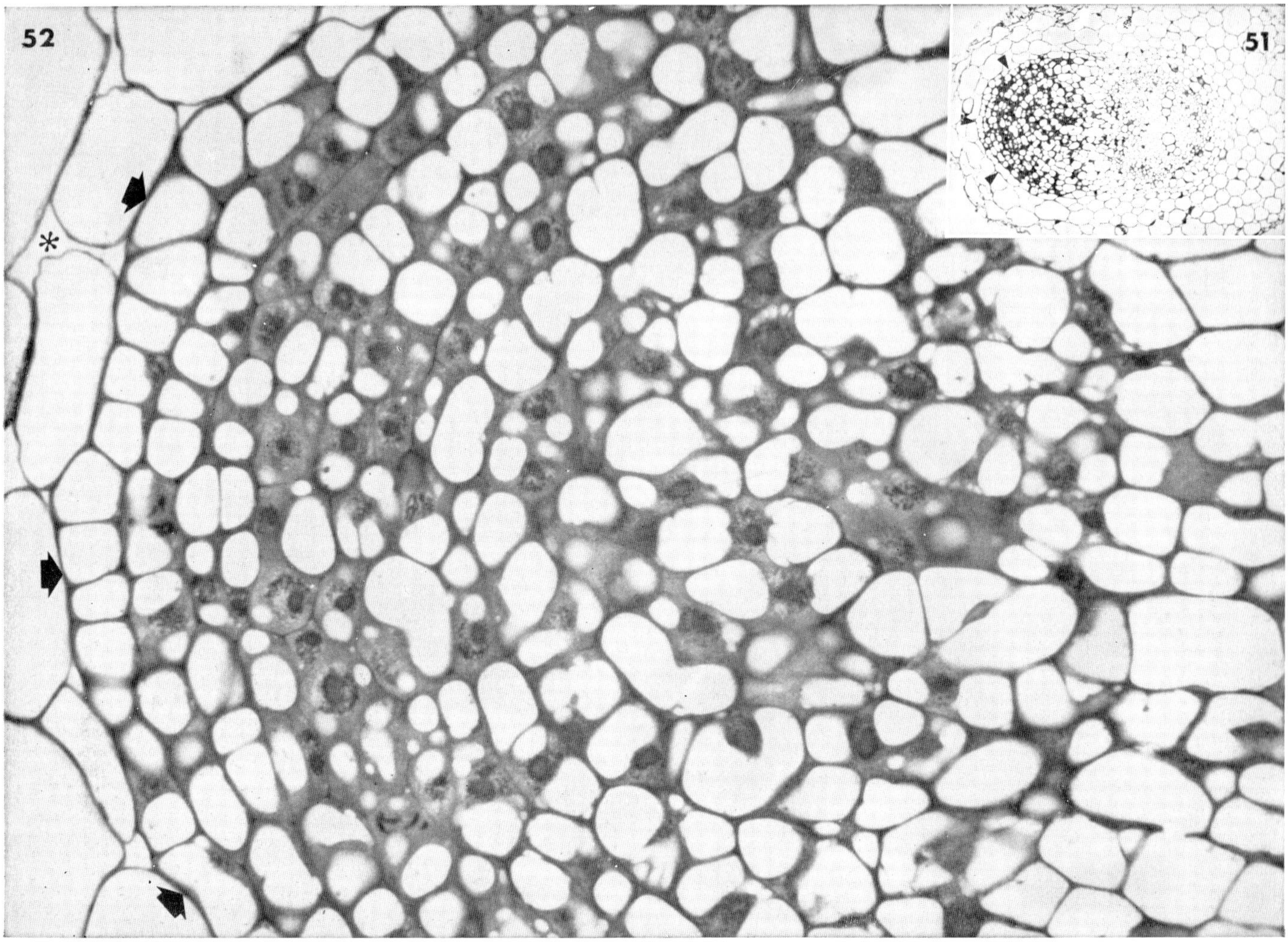

51-52 Lateral root initiation in pea.

51 Transverse section of a main root showing a well-developed lateral root primordium. Toluidine blue. × 74.

52 Part of the lateral root primordium shown in Figure 51 at higher magnification. Note the complete absence of secretory cells over the surface of the primordium (arrows). The cortical cells appear simply to "come apart" (asterisk). Toluidine blue. × 680.

53-54 Tissues of a root with secondary thickening (from a blackberry, *Rubus fruticosus*). Both stained with toluidine blue.

53 Part of the root in transverse section, showing the secondary xylem (X) and phloem (Ph), proliferated pericycle (Pe), cork (Co), and cork cambium (CoC). × 45.

54 Part of Figure 53 in greater detail showing the secondary xylem (X) and phloem (Ph). Ray cells (formed from initials in the cambial zone (CZ) which elongate preferentially parallel to the radius of the root) accumulate starch in the phloem rays (PR) and tannin in the xylem rays (XR). P, plastid; Pi, pits; TE, tracheary element; TC, tannin cell; XF, xylem fiber. × 430.

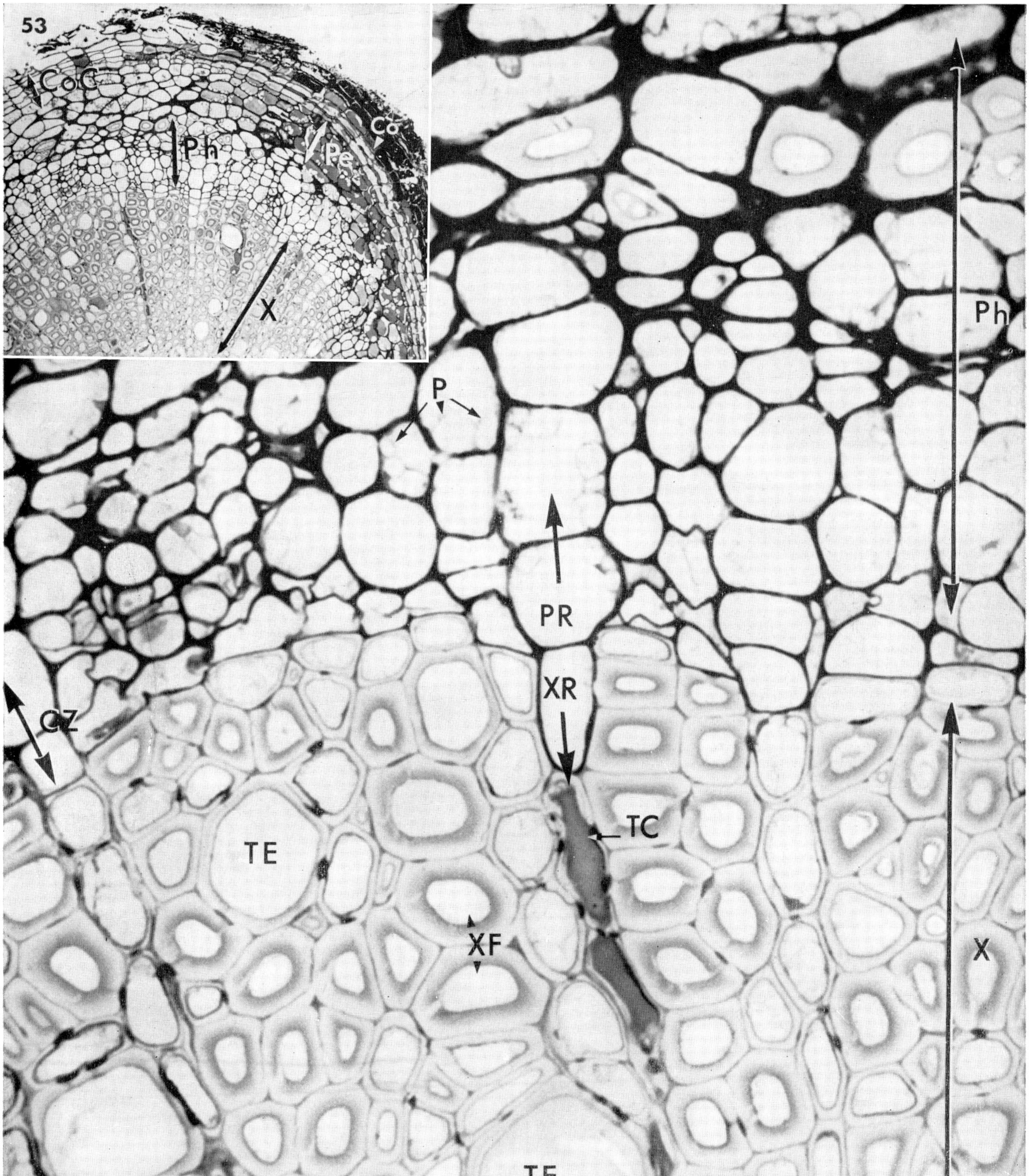
53
CoC
Ph
Pe
Co
X
P
PR
XR
CZ
TC
TE
XF
X
Ph
Pi
TE
54

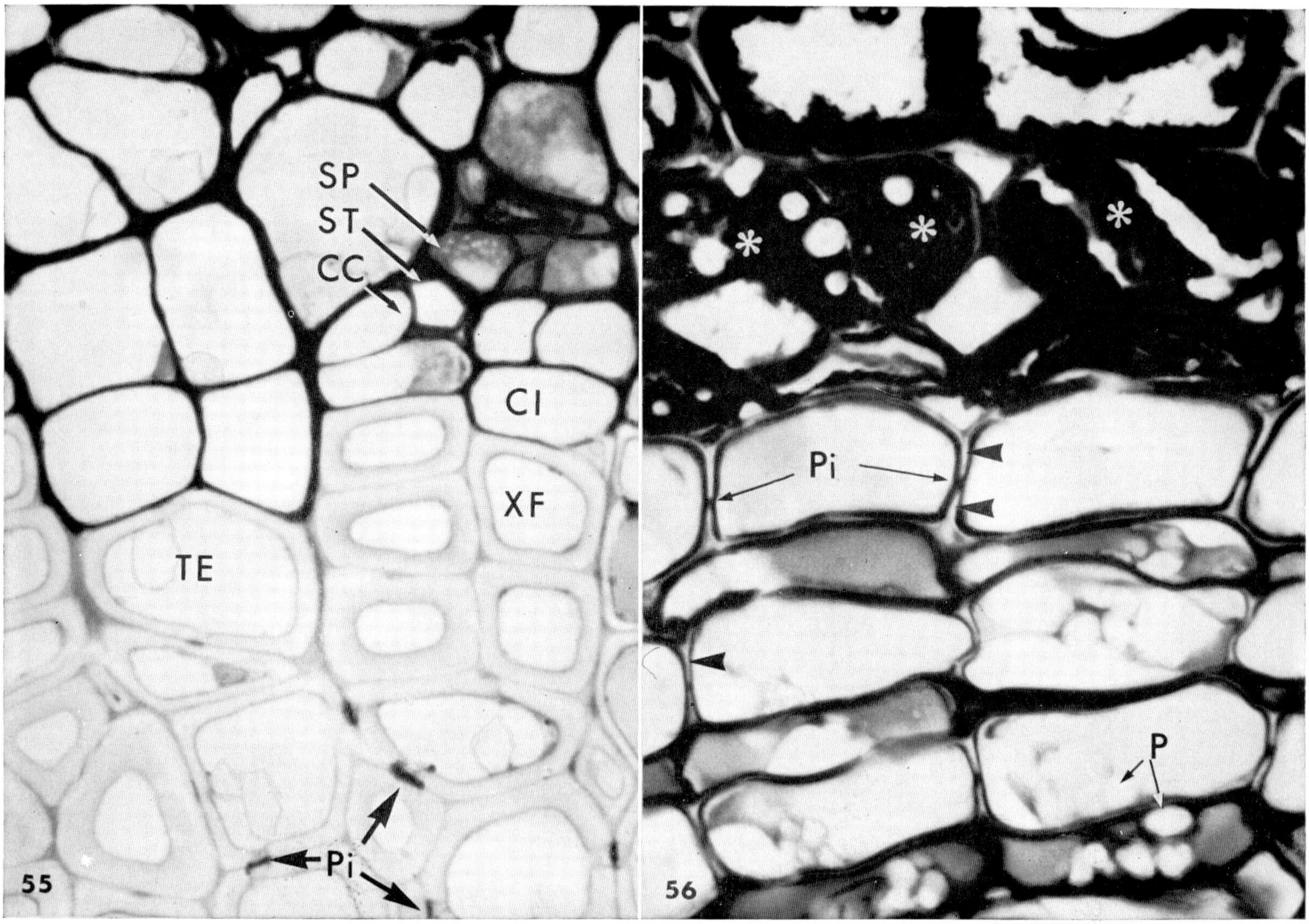

55 The cell types of the secondary xylem and phloem, from the same root as that shown in Figure 53. Note the tracheary elements (TE) and fibers (XF) of the xylem, starch-rich parenchyma cells, sieve tubes (ST), sieve plate (SP), and companion cells (CC) of the phloem. CI, cambial initial, Pi, pits. × 940.

56 Cork cambium and cork from the outer layers of the root shown in Figure 53. These layers derive from the proliferated pericycle of the primary root. Note the lipid-rich walls (with pits, Pi) of these layers which fail to stain with toluidine blue (arrows), as did the lipid-rich walls of endodermal cells (Figure 45). The cork cells accumulate an abundance of tannins (asterisks). P, Plastid. Both stained with toluidine blue. × 940.

The Initiation and Early Development of Lateral Roots

Roots of grasses, especially those of certain varieties of maize, are excellent material in which to study the origin and development of lateral roots. In these plants, the cortex of the root can be peeled away easily with fine forceps, exposing *lateral root primordia* in various stages of development (Figure 47).

It is often stated in texts that lateral roots arise endogenously within the stele. In maize, this is true only for the very first cell divisions that mark the initiation of a new primordium. These first divisions occur parallel to the long axis of the root within the pericycle (Figure 48). A zone of growth and division spreads rapidly from this initial site and soon encompasses cells both of the stele and cortex (Figures 49 and 50). Indeed, the original mature pattern of cells within the cortex and stele is completely altered and by the time the new primordium has reached the root surface, the original pattern is scarcely visible at the site of the primordium.

The origin of lateral roots in dicotyledons is somewhat different from that in grasses. In most dicotyledons, the first divisions within the pericycle occur opposite or close to a protoxylem pole, whereas in grasses they occur opposite a protophloem pole (Figures 48 and 49). Secondly, the surface layer of cells of a differentiating root primordium in maize develops into an epithelium of polarized cells (Figure 50). These cells are coated with a thick layer of extracellular material, and material with a similar staining behavior occurs in the apical pole of these cells (Figure 50, inset). The cortical cells of the parent root collapse completely wherever they contact this epithelial surface, whereas cells not in contact appear to remain unchanged (Figure 50). In contrast, the cells at the surface of a developing root primordium in pea (Figures 51 and 52, arrows) are not polarized and in no sense resemble a secretory epithelium. The cortical cells of the parent pea root appear simply to come apart without collapse, as the primordium emerges (Figure 52, asterisk). It is not known whether this is a constant difference between grasses (or other monocotyledons) and dicotyledons.

An overwhelming body of physiological evidence shows that auxins (such as indole acetic acid) are directly involved in the initiation of lateral roots. The initiation of roots at cut stems, production of lateral roots by cultured whole roots or root segments, and the initiation of roots from callus cultures are all regulated by the concentration of auxin at the site of initiation (*3.7*). Although auxin plays a dominant role in root initiation, it seems certain that it is not the only hormone regulating this process. Studies on isolated segments of *Convolvulus* roots have shown that whether a primordium initiated in a root segment develops into a shoot or a root depends upon the concentration of cytokinins in the medium (*3.8*). It is well known that lateral root initiation is inhibited by the presence of a rapidly growing root apex. Decapitation of a main root promptly leads to the formation of laterals close to the cut surface, even though primordia may have been absent for 20 cm. from the apex. These facts have led to the suggestion that the root tip is a source of cytokinins, which inhibit lateral root initiation for some distance behind the tip. Analysis of the xylem contents from isolated root systems has shown that cytokinins are present in it but direct proof that these molecules are produced in the root tip is still wanting.

Secondary Thickening of the Root

The roots of most woody plants develop a cambium and secondary xylem and phloem (Figure 53). The cambial initials arise within the stele at the inner edges of the phloem while the metaxylem elements are still differentiating (Figure 46). This wave of oriented cell division and growth spreads towards the xylem poles and forms a complete cambial zone when the pericycle at the xylem poles also enters division. Eventually, this cambial zone forms a complete cylinder (Figure 53) because secondary xylem differentiates first in the region that underlies the primary phloem. The sieve elements of the primary phloem are crushed and many of the remaining cells form fibers. The cell types of the secondary phloem and xylem are illustrated in Figures 54 and 55.

Divisions of the pericycle continue and spread circumferentially from the xylem poles. A second cambium, the *phellogen* (or *cork cambium*), develops within this proliferated pericycle. The phellogen divides periclinally to produce the *periderm* tissue. Cells of the periderm produced on the outside of the phellogen have lipid-rich walls, which fail to stain with cationic dyes (Figure 56, arrow; cf. endodermal cell walls, Figure 45), and frequently these cells are filled with tannins. The original cortex and epidermis are crushed by the expanding periderm and sloughed off. The cork cells then provide an effective barrier to pathogen invasion.

GENERAL REFERENCES

Clowes, F. A. L. *Apical Meristems.* Oxford: Blackwell Scientific Publications, 1961.

Esau, K. *Plant Anatomy.* New York: John Wiley & Sons, 1965, Chap. 17, 481–538.

LEOPOLD, A. C. *Auxins and Plant Growth.* Berkeley: University of California Press, 1963.

TORREY, J. G. *Development in Flowering Plants.* New York: The Macmillan Co., 1967, Chap. 5, 78–93.

REVIEWS AND RESEARCH PAPERS

3.1 CARR, D. J. "Metabolic and hormonal regulation of growth and development," in *Trends in Plant Morphogenesis*, ed. E. G. Cutter. London: Longmans, Green and Co., Ltd., 1966.

3.2 CORMACK, R. G. H. "The development of root hairs in angiosperms," *Bot. Rev.*, 15 (1949), 583–612.

3.3 CLOWES, F. A. L. "Apical meristems of roots," *Biol. Revs.*, 34 (1959), 501–529.

3.4 CLOWES, F. A. L. "The quiescent centre in meristems and its behaviour after irradiation," in *Meristems and Differentiation*, Brookhaven Symposium in Biology, No. 16, 1964.

3.5 WHITE, P. R. "Nutritional requirements of isolated plant tissues and organs," *Ann. Rev. Plant Phys.*, 2 (1951), 231–244.

3.6 BUTCHER, D. N., AND H. E. STREET. "Excised root culture," *Bot. Rev.*, 30 (1964), 513–586.

3.7 TORREY, J. G. "Physiological bases of organization and development in the root," in *Encyclopaedia of Plant Physiology*, XV, pt. 1, ed. W. Ruhland. Berlin: Springer-Verlag, 1965.

3.8 BONNETT, H. T., AND J. G. TORREY. "Comparative anatomy of endogenous bud and lateral root formation in *Convolvulus arvensis* roots cultured *in vitro*," *Amer. J. Bot.*, 53 (1966), 496–507.

57 Opening buds of *Syringa vulgaris* (lilac). × 4.

4

The Shoot Apex and Leaf Initiation

SHOOTS ELONGATE at their apices. At no time is this apical growth habit seen more clearly than in the spring in temperate climates. Shoot apices that have spent the winter as dormant buds suddenly elongate, burst from their confining bud scales, and add their leaves and stem tissues to the distal ends of their supporting twigs. Such apical growth has just been initiated by the two young shoots that have formed near the tip of the lilac twig shown in Figure 57. The new leaves that appear above the bud scales (Figure 57) and other leaves that are still inside (Figure 58) or as yet unformed, along with new stem tissue, will be added to these twigs. Thus, by the end of the summer, the apices shown in Figure 57 will each be at the end of a new piece of woody twig, the cumulative product of their season's growth.

A small region of continued juvenility is present at the extreme apex of a vegetative shoot (Figures 58–62). Behind this tip there is laid out in space an ontogenetic sequence from the undifferentiated cells of the apex to the tissues of the mature shoot. Throughout its life a shoot apex is pushed continuously away from the main body of the plant by the extension of some of its newly formed cells. These cells subsequently pass through a subapical developmental sequence and are added to the mature shoot tissues. Besides adding to the mature cells of the shoot, the apical region also maintains itself by the production of new undifferentiated cells.

Despite the great variation in the form of the adult tissues of leafy plants, their shoot apices are surprisingly similar in size and morphology. All have a central zone of meristematic cells. This zone is relatively small, ranging in its largest diameter (measured at the level where leaf primordia first arise) from approximately 100μ to 500μ in most angiosperms and gymnosperms up to about 3,000μ for the largest apices, those of some of the ferns and cycads. Furthermore, all shoot apices produce leaf primordia on the flanks of the meristematic zone. No other projections form on the smooth periphery of the meristem above the level of the youngest leaf primordia (Figures 58–62).

With the exception of the single, very large apical cell that lies at the surface in the center of the apical meristem of mosses, horse tails, and many ferns, the cells that lie above the level of the first leaf primordia of apical meristems are small and isodiametric. In thin sections (Figure 60) these cells can be seen to contain many small vacuoles and their nuclei to occupy a relatively large proportion of the volume of each cell.

The shape of the apical meristem above the level of the youngest primordium does differ in different plants, and the apex may range from being almost flat to forming a high thimblelike dome (Figures 58–62).

The Formation of Leaves

The production of leaf primordia is the most characteristic feature of the apical meristems of vegetative shoots: indeed, leaf primordia never form at any other site. The initiation and early development of a leaf primordium appear to be quite similar in all species. They are well displayed in the aquatic flowering plant, *Hippuris*, or mare's tail, because the shoot apex of this plant is unusually long. It includes many whorls of leaf primordia and immature leaves, so that many stages in leaf development can be seen in a single longitudinal section through the apex (Figures 61 and 62).

Some of the stages in the initiation and development of a leaf primordium in *Hippuris* (*4.1*) are illustrated in Figures 63 to 66. The first sign of the initiation of a primordium is an increase in baso-

58 **Longitudinal section through an opening lilac bud showing the shoot apex (arrows) and portions of several immature leaves (IL). PAS/toluidine blue. × 88. (Section courtesy Mr. K. R. Brasch and Mrs. M. Brasch.)**

59 **Longitudinal section through a shoot apex of *Tropaeolum* spp. (nasturtium) showing the apical dome (arrows), portions of immature leaves (IL), and an axillary bud (AB). PAS/toluidine blue. × 88.**

60 **Higher magnification of the apical dome region shown in Figure 59. The arrow indicates a developing leaf primordium. PAS/toluidine blue. × 690. (Section courtesy Mr. R. G. Fulcher and Mr. H. B. Younghusband.)**

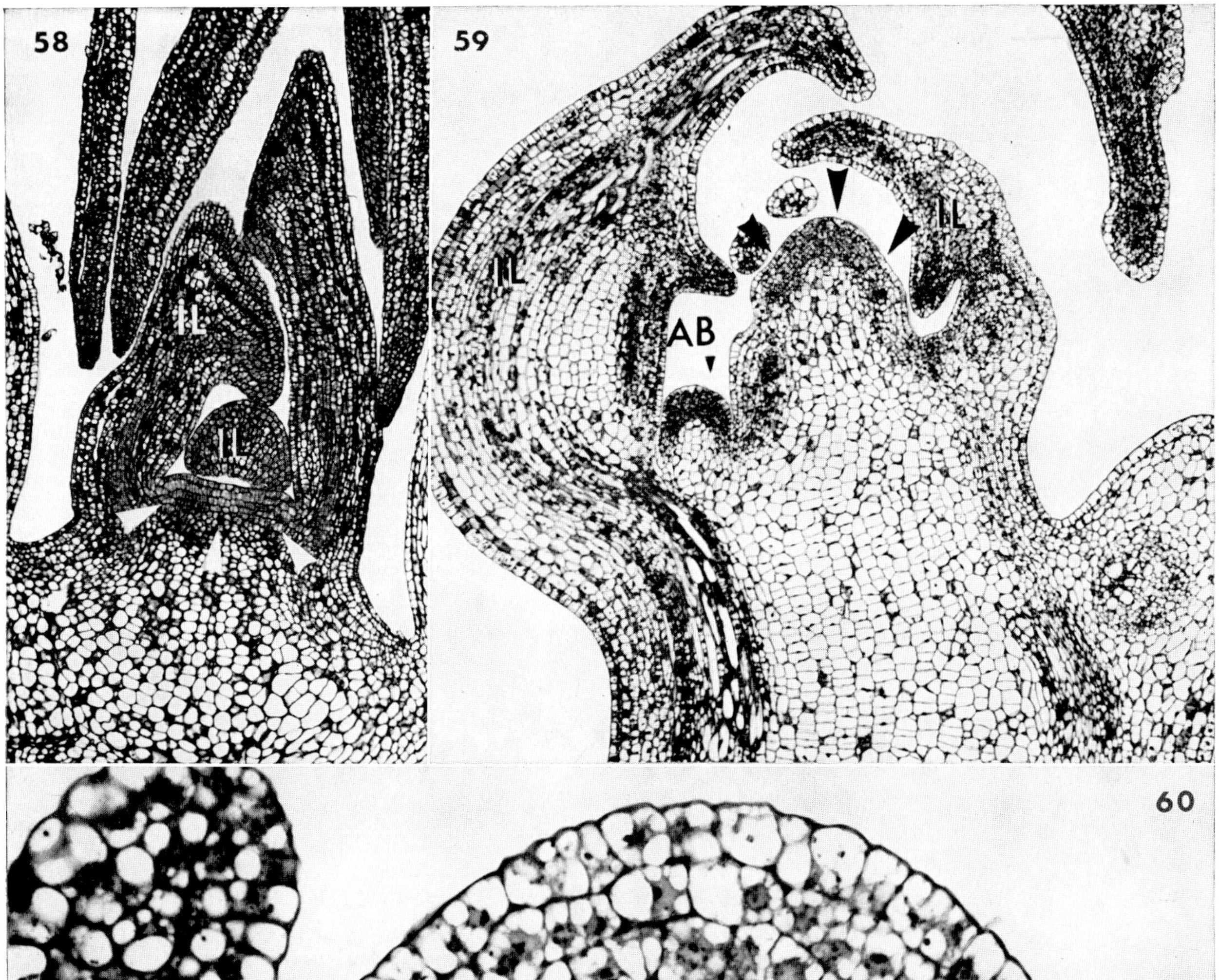
58
IL
IL
59
IL
IL
AB
60
IL

philia and a slight enlargement of a small group of cells, which lie in the subepidermal layer of the apical dome. Subsequently, some of these cells divide periclinally (Figure 63), and the enlargement of their daughter cells produces a small bump on the surface of the apex. Anticlinal divisions of the cells of the outer layer of the apex (Figure 63) allow this layer to extend and keep pace with the developing bulge. Similar division and enlargement of the cells of the inner group and the cells of the peripheral layer continue (Figure 64) until the primordial bulge is approximately 50μ long (Figure 65). At about this stage the cells of the peripheral layer that covers the primordium differentiate to form the epidermal cells of the leaf. The inner cells differentiate into mesophyll parenchyma and provascular tissue, and the primordium becomes an immature leaf (Figure 66).

The sites at which leaf primordia form are of prime importance in determining the morphology of the shoot. Not only are the positions of the primordia reflected directly in the arrangement of the mature leaves, but the nodes and internodes of the stem also differentiate in relation to the position of the leaves. A disc of nodal tissue always forms across the stem at the level of each leaf primordium, and an internode develops between each node. The correlation between the position of nodes and that of young leaves is particularly evident in longitudinal sections through the shoot apex of *Hippuris* (Figures 61 and 62). Portions of one or two primordia of each whorl are seen in each section, and the nodal and internodal tissues of this plant can be easily distinguished quite close to the shoot apex.

The distinction between young nodes and internodes is particularly clear outside the stele (Figures 61 and 62). The internodal tissues of the cortex of such water plants (see also Figure 1) develop large intercellular spaces, whereas the node remains as a thin disc of closely packed cells, some of which differentiate into provascular tissue.

Leaf primordia arise in an orderly fashion on the shoot apex. The positions of successive primordia usually bear a regular mathematical relationship to each other, frequently lying along one or more spirals. Because of this regular arrangement it is possible to predict where new primordia will form on the apical dome. This fact has greatly facilitated experimental surgery on shoot apices (*4.2, 4.3, 4.4, 4.5, 4.6, 4.7*). By selectively damaging the meristematic dome and/or the young leaf primordia, it has been possible to evaluate some of the factors that control the position and develop-

61 **Median longitudinal section through the apex of an aerial shoot of *Hippuris vulgaris* (mare's tail) showing the apical dome (arrows), leaf primordia (LP), and immature leaves (IL). Intercellular space (IS). Toluidine blue/acid fuchsin. × 160.**

62 **A similar section to that in Figure 61 but of an aquatic shoot. Intercellular space (IS). Toluidine blue/acid fuchsin. × 160.**

63-66 **Early stages in the development of an aquatic leaf in *Hippuris*. All sections stained with toluidine blue/acid fuchsin.**

63 **Longitudinal section through the flank of an apical dome. The section passes through the center of a very young leaf primordium showing a recent periclinal division in the subepidermal layer (the new wall can be seen at the tip of the open arrow) and the anaphase of an anticlinal division in the epidermis (solid arrows). × 1400.**

64 **Longitudinal section through an older leaf primordium which is now bulging out from the apical dome. × 1400.**

65 **Longitudinal section of a still older leaf primordium. × 1400.**

66 **Longitudinal section through portions of several immature leaves that have started to differentiate epidermal and mesophyll tissue. The section shows an epidermal gland differentiated by one of the leaves. × 440.**

67 **Transverse section of mature aerial (left) and aquatic (right) leaves of *Hippuris*. PAS/toluidine blue. × 125.**

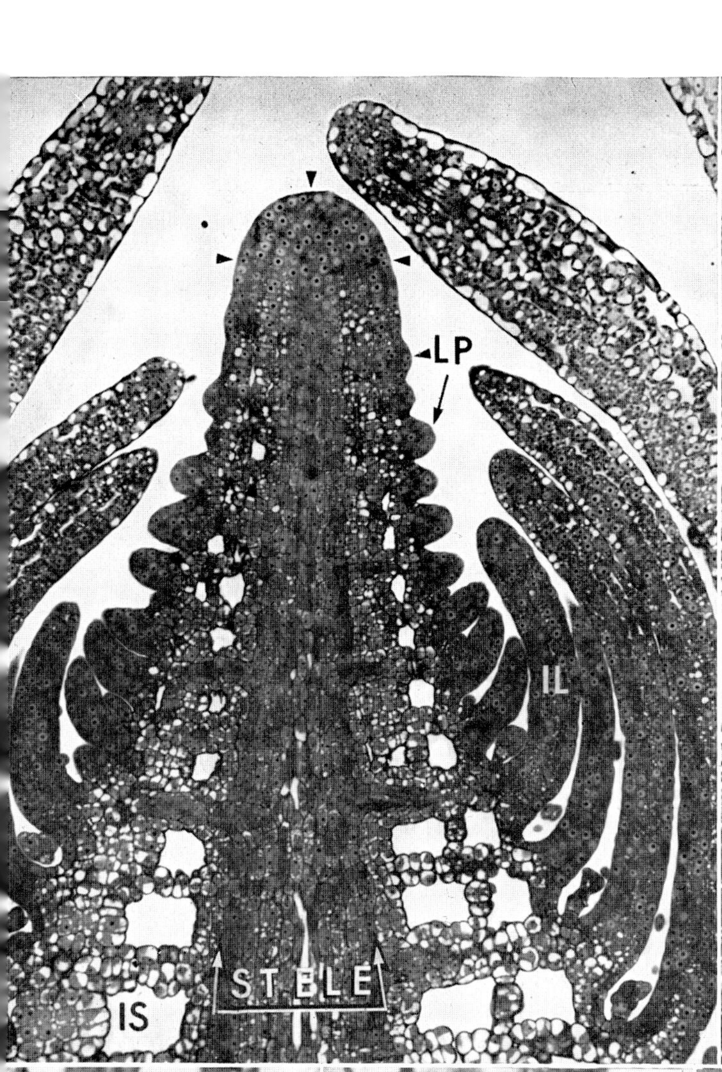
LP
IL
IS
STELE

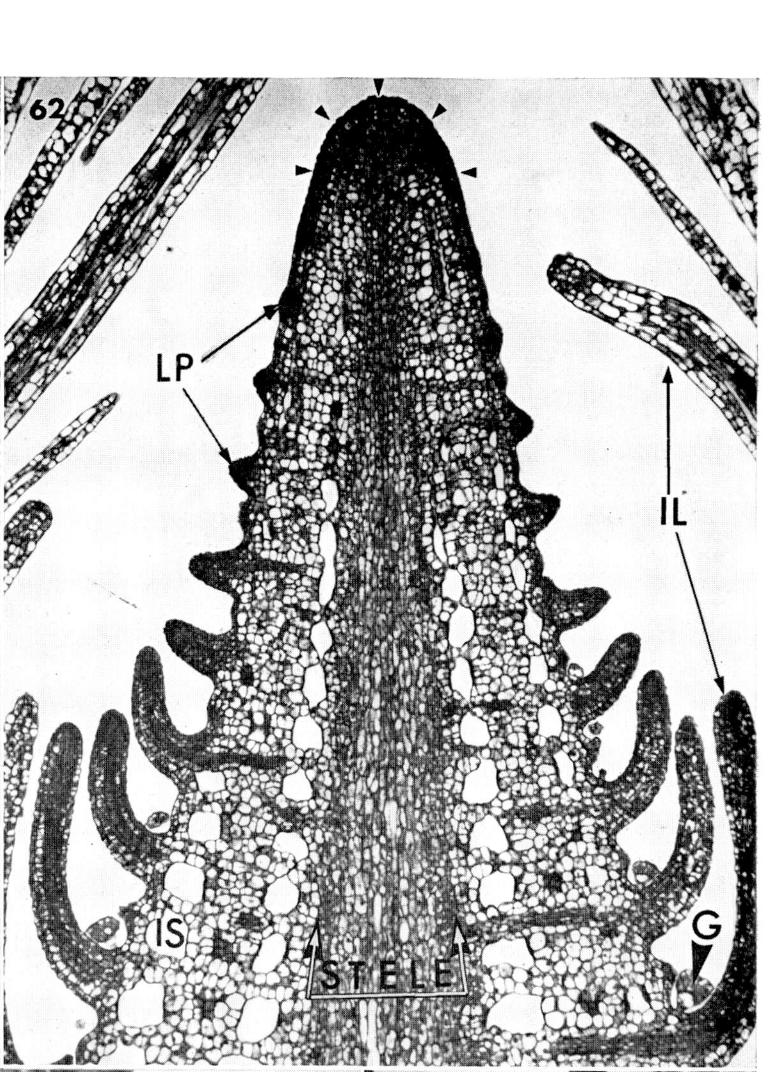
62
LP
IL
IS
STELE
G

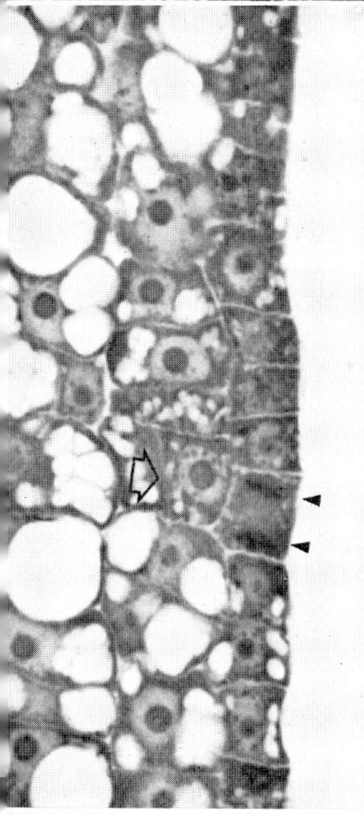

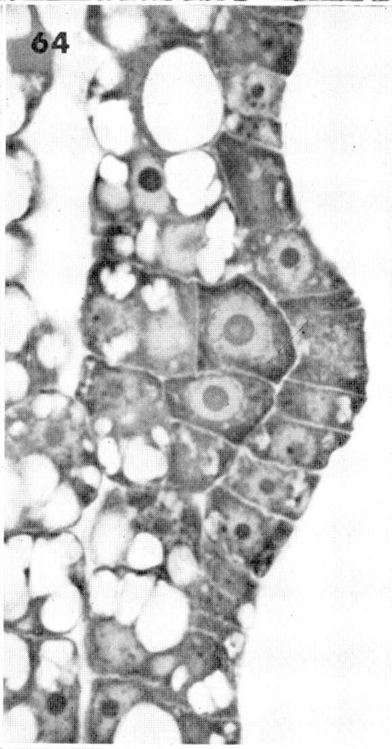
64

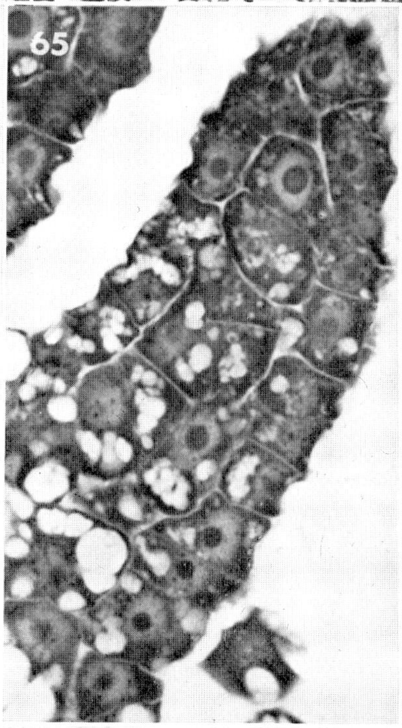
65

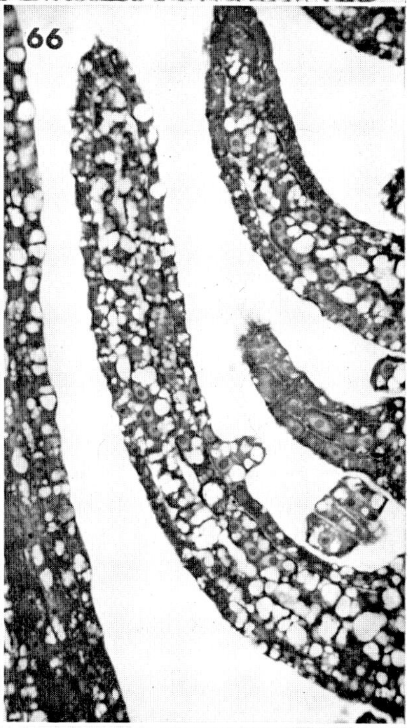
66

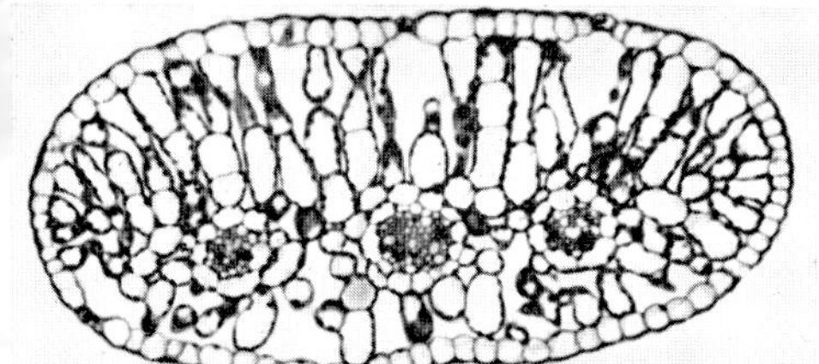

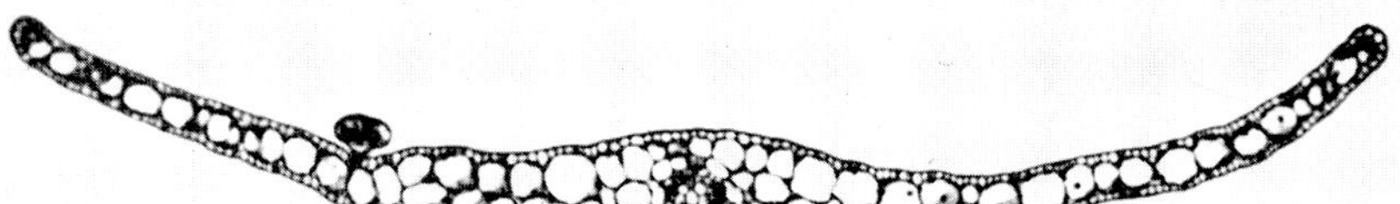

ment of new primordia. Results of these and other experiments on shoot apices have led investigators to postulate that a new primordium arises in the "first available space" on the apical dome (*4.7, 4.8*). There is more involved than just the presence of a physical space for the new primordium. Much of the experimental work has indicated that both the intact meristematic dome and the young primordia inhibit the initiation of new primordia. The first space available for the formation of a new primordium appears to be the area on the periphery of the dome where these inhibitory effects first reach a nonlimiting level. The exact nature of the inhibition is not known but it is almost certainly hormone mediated.

Changes in Shoot Apices

During its lifetime a shoot apex may undergo changes in its morphology and in the nature of the tissues it produces. Some of these changes occur in the normal ontogeny of an apex as discussed later in this section, but others are imposed directly by variations in the external environment of the plant (*4.9*).

One of the classic examples of the influence of the environment on shoot apices occurs in amphibious plants such as *Hippuris* (*4.10*). Here the shoots begin growth under water but reach the surface and continue to grow in the air. Comparisons of sections through aquatic (Figure 62) and aerial (Figure 61) apices of *Hippurus* show that the transition from the aquatic to the aerial environment results in a shortening and thickening of the apical dome. At the same time the leaf primordia of aerial apices mature at a level closer to the apical dome than is the case in corresponding aquatic apices. Furthermore, although the initiation and early development of the leaf primordia of both types of apex appear to be similar, their subsequent differentiation is quite different (Figure 67). Primordia that differentiate in the air form short, thick leaves with well-developed mesophyll, net venation, and stomata, whereas those that differentiate under water form long, narrow leaves with little mesophyll, linear venation, and no stomata.

Another example of the influence of the environment on shoot apices is the induction of their dormancy by long nights and low temperatures (*4.11*) (see Chapter 6). When dormancy is induced, shoot apices frequently change shape, and their internodes fail to elongate. The developmental sequence below the apex is slowed down and many immature leaves may accumulate. With the onset of dormancy some of the older primordia may differentiate into bud scales (see Chapter 6).

The most profound change that occurs in shoot apices is their transition from the vegetative to the floral state. Frequently this transition is stimulated by the environment, being induced in some plants by long nights and in other plants by short nights. When flowering is induced, the apex loses its capacity to replenish its content of meristematic cells and the whole of the apical dome becomes involved in the production of the floral organs (see Chapter 8). The apex loses its potential for continued elongation, that is, the growth of the shoot becomes determinate.

The Autonomy of Shoot Apices

A number of investigators have shown that the vegetative shoot apices of a variety of angiosperms and ferns can be isolated and cultured on a relatively simple medium (*4.12, 4.13*). In all cases, as long as the isolated apex includes a few young leaf primordia, it will continue to maintain itself and to produce leaves and stem tissue of normal morphology regardless of its isolation from the mature plant. It becomes clear from this work with cultures that the organization of a shoot apex is inherent in the apex and not dictated by the mature tissues of the plant. However, the degree of autonomy of shoot apices varies in different types of plants. In the case of angiosperms it has been found that an isolated apex will not develop into a shoot unless at least one leaf primordium is present (*4.12*). On the other hand, in ferns apical tips isolated *above* the level of the youngest leaf primordia will develop new leaf primordia and form normal shoots in culture (*4.13*).

The Origin of Shoot Apices

Shoot apices normally originate in one of two ways; either they develop at one end of the axis of an embryo (Figures 148, 149, and 151) or they arise as lateral buds in the axils of leaves (Figure 96).

The shoot apex of a dormant embryo usually consists of little more than a dome of meristematic cells (Figures 148, 149), but in some species a few leaf primordia and immature leaves may be produced before the embryo becomes dormant (Figure 151). The shoot apices of both embryos and lateral buds are smaller and often flatter in shape (compare Figures 96 and 97; see also Figures 149 and 151) than those of the corresponding mature shoots. Their mature size and shape is reached gradually during the growth of the young shoot. This ontogeny of the apex is accompanied frequently by changes in the arrangement and morphology of the leaves it produces.

Although new shoot apices normally form in embryos or axillary buds of shoots, such an origin is apparently not obligatory as they have been shown to originate in quite different tissues under experimental conditions. For example, shoot apices can form in tissue cultures of a wide variety of plant species (*4.14, 4.15*). Cells in tissue cultures usually remain large and vacuolated, and they divide and grow to form aggregations of parenchyma cells termed *callus tissue.* A variety of hormonal treatments will initiate the formation of nodules of highly meristematic cells in such callus tissue, and these nodules may be stimulated to organize into shoot apices, which subsequently form normal leafy shoots.

Perhaps the most surprising place in which shoot apices have been experimentally induced to form is in isolated root segments in culture. Under some conditions, shoot apices develop within groups of meristematic cells that proliferate from the pericycle and surrounding tissues of cultured roots. In the roots of intact plants, such groups of meristematic cells usually organize into apices of lateral roots (see Chapter 3), but in culture, they can be forced to form either root or shoot apices, depending upon the conditions of culture (*3.7*).

What dictates the formation of a shoot apex? The experimental evidence available shows that under the right conditions (as yet only vaguely defined) groups of meristematic cells of widely different origin will organize into shoot apices characteristic of the species involved. The ultimate source of the information that controls the organization of a shoot apex lies, of course, in the genetic material of the component cells. How this information is translated to direct the development of a complex, self-maintaining apex remains a problem for the future.

GENERAL REFERENCES

CLOWES, F. A. L. *Apical Meristems.* Oxford: Blackwell Scientific Publications, 1961.

ESAU, K. *Plant Anatomy*, 2d ed., Chaps. 4 and 5. New York: John Wiley & Sons, 1965.

ROMBERGER, J. A. "Meristems, Growth and Development in Woody Plants." *U.S. Department of Agriculture Technical Bulletin*, No. 1293, 1963.

SINNOTT, E. W. *Plant Morphogenesis*, Chap. 4. New York: McGraw-Hill Book Co., 1960.

WARDLAW, C. W. *Morphogenesis in Plants*, Chaps. 3–5. London: Methuen & Co., Ltd., 1952.

———. *Organization and Evolution in Plants*, Chap. 10. New York: Longmans, Green & Co., 1965.

REVIEWS AND RESEARCH PAPERS

4.1 MCCULLY, M. E., AND H. M. DALE. "Variations in leaf number in *Hippuris*: A study of whorled phyllotaxis," *Can. J. Bot.*, 39 (1961), 611–625.

4.2 CUTTER, E. G. "Recent experimental studies of the shoot apex and shoot morphogenesis," *Bot. Rev.*, 31 (1965), 7–113.

4.3 WETMORE, R. H., AND C. W. WARDLAW. "Experimental morphogenesis in vascular plants," *Ann. Rev. Plant Phys.*, 2 (1951), 269–92.

4.4 SNOW, M., AND R. SNOW. "Experiments on phyllotaxis I. The effect of isolating a primordium," *Phil. Trans. Roy. Soc. B.*, 221 (1931), 1–43.

4.5 SNOW, M., AND R. SNOW. "Experiments on phyllotaxis II. The effect of displacing a primordium," *Phil. Trans. Roy. Soc. B.*, 222 (1933), 353–400.

4.6 SNOW, M., AND R. SNOW. "Experiments on phyllotaxis III. Diagonal splits through decussate apices," *Phil. Trans. Roy. Soc. B.*, 225 (1935), 63–94.

4.7 WARDLAW, C. W. "Phyllotaxis and organogenesis in ferns," *Nature*, 164 (1949), 167–169.

4.8 SNOW, R. "Problems of phyllotaxis and leaf determination," *Endeavour*, 14 (1955), 190–199.

4.9 ALLSOPP, A. "Heteroblastic development in cormophytes". In *Encyclopaedia of Plant Physiology*, XV, ed. W. Ruhland. Berlin: Springer-Verlag, 1965.

4.10 MCCULLY, M. E., AND H. M. DALE. "Heterophylly in *Hippuris*: A problem in identification," *Can. J. Bot.*, 39 (1961), 1099–1116.

4.11 WAREING, P. F. "Photoperiodism in woody plants," *Ann. Rev. Plant Phys.*, 7 (1953), 191–214.

4.12 BALL, E. "Development in sterile culture of stem tips of *Tropaeolum majus L.* and of *Lupinus albus L.*, *Amer. J. Bot.*, 33 (1946), 301–318.

4.13 WETMORE, R. H. "The use of *in vitro* cultures in the investigation of growth and differentiation in vascular plants," *Brookhaven Sympos. Biol.*, 6 (1954), 22–40.

4.14 SKOOG, F., AND C. O. MILLER. "Chemical regulation of growth and organ formation in plant tissues cultured *in vitro*," *Sympos. Soc. Exp. Biol.*, 11 (1957), 118–131.

4.15 EARLE, E. D., AND J. G. TORREY. "Morphogenesis in cell colonies grown from *Convolvulus* cell suspensions plated on synthetic media," *Amer. J. Bot.*, 52 (1965), 891–899.

68 Simple hairs and "stinging hairs" of the leaf of *Urtica* spp. (stinging nettle). × 12.

5

The Leaf

IN MOST green plants (but by no means in all) leaves contain the bulk of the photosynthetic tissue. The external morphology of leaves is among the most variable features of plants, and yet the leaves of all vascular plants have certain important feaures in common. These basic features are sometimes difficult to see because variation in the arrangement of the tissue systems tends to obscure the physiologically important units.

A group of parenchyma cells that contain chloroplasts is undoubtedly the most fundamental unit of structure and function in a leaf, and indeed in any photosynthetic tissue of a vascular plant. These units are exposed to the air and supplied with a vascular system. From this vascular system they receive water and solutes (inorganic and organic), and to it they deliver a limited range of photosynthetic products for distribution to the nonphotosynthetic tissues of the plant. These units cannot function efficiently without protection and support. All leaves of vascular plants consist of arrays of these photosynthetic units, protected and partially regulated by epidermal layers and supported in the radiation flux by *sclerenchyma* and *collenchyma*.

The *precise arrangement* of these units and the distribution of protective and supportive tissue systems is as variable as leaf morphology itself. Only two examples are shown here (Figures 69, 70, 71 and 72). Each of these examples illustrates, in a general way, the strikingly different patterns that characterize the dicotyledonous and monocotyledonous leaf. Clearly, various groups of plants have developed a variety of successful solutions to the same general problems posed by their lives as terrestrial photosynthetic autotrophs.

Tissue differentiation in leaves is not treated in this book. However, it must be remembered that the complex patterns shown in Figures 69 and 70 develop solely as a result of differential rates of growth and carefully regulated sequences of cellular differentiation (*5.1*). Except for a certain amount of intrusive growth by sclerenchyma cells, cell migration, so characteristic of development in animal tissues, does not occur in plants. Some displacement does occur between cells during the formation of gas spaces, but this is always accompanied by localized rupture of the extracellular material (*5.2*).

In this section attention is directed towards the chloroplast-rich parenchyma and the epidermal layers because both of these tissues have functions peculiar to photosynthetic organs. On the other hand, the vascular tissues of leaves are not known to have major functions different from those which they carry out in the stem (see Chapter 7).

The Chloroplast

If one tears a strip from a living mature green leaf and examines the vacuolated chlorophyll-containing *mesophyll* cells at high magnification in the light microscope, it is easy to confirm that all of the green pigment is contained within discrete bodies. These *chloroplasts* (2–5μ in maximum dimension) are usually confined to the parietal cytoplasm, but they may be moved around the cell by cytoplasmic streaming and they show marked amoeboid movement in the living cell. Commonly, one may distinguish regions within each plastid, the *grana*, which are specially rich in chlorophyll, and clear refractile granules of starch, especially if the leaf has been photosynthesizing for a few hours. The pigments are usually extracted during dehydration when leaf tissues are prepared for histologic study, but the starch is retained and the bulk of the chloroplast stains strongly with cationic dyes (Figure 73).

In the electron microscope, a wealth of additional structure can be discerned (Figure 74). Like mitochondria (Figure 10), all plastids are bounded by two membranes, and in a few places the inner membrane forms shallow invaginations. Much of the volume of the chloroplast is made up of a matrix of moderate electron density, the *stroma*. The most striking feature of the chloroplast, however, is the internal membrane system. This consists of a number of membrane-rich stacks, the grana, that are connected to one another by one or more lamellae, the *stroma lamellae*. The three-dimensional structure of this membrane system in the mature chloroplast is very variable and quite controversial, but the following interpretation has good support. It has been suggested that outgrowths from the stroma lamellae form flattened membrane-bounded pouches (*5.3*). Part of the outer surface of these pouches is closely appressed to the outer surface of the parent lamella, whereas another part may be appressed either directly to a different stroma lamella or to a similar pouch formed from another stroma lamella. A careful scrutiny of Figure 74 and inset shows that the dark lines of the internal membrane system are indeed just such

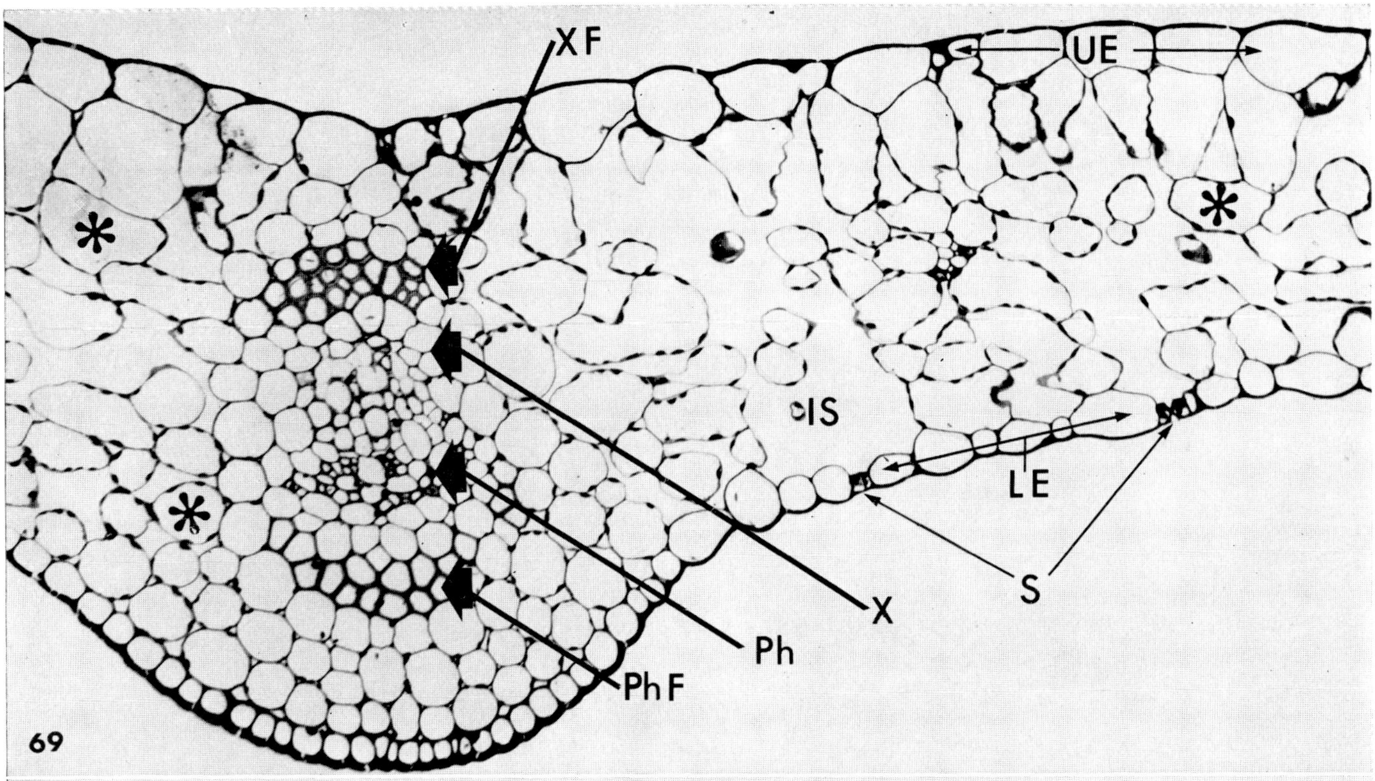

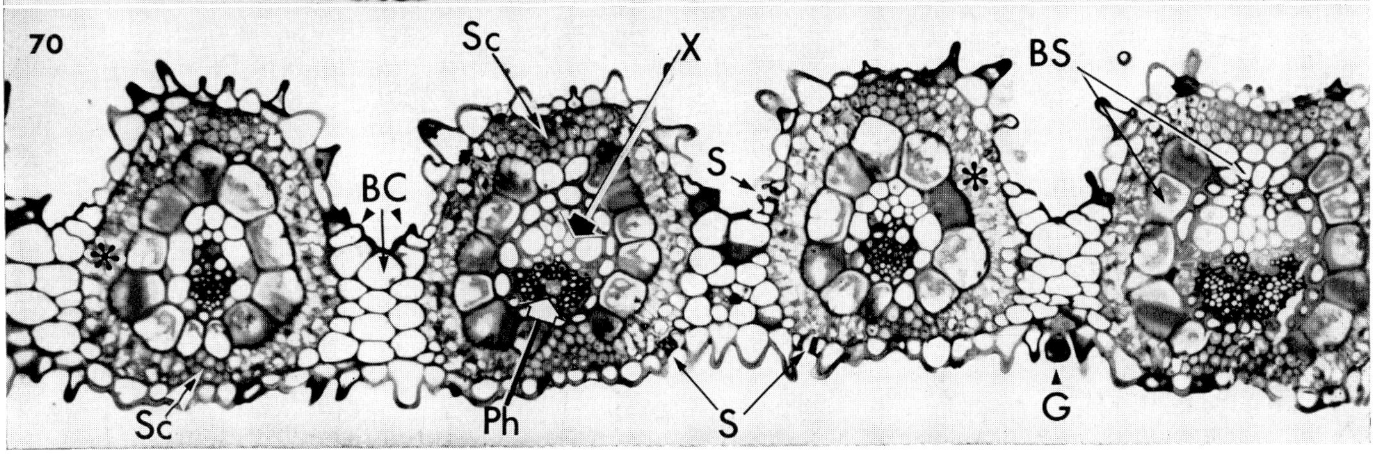

69 **Transverse section of pea leaf, showing the typical arrangement of tissues in a dicotyledonous leaf. Note the intercellular spaces (IS) between the chlorenchyma cells (asterisks). LE, lower epidermis; Ph, phloem; PhF, phloem fibers; S, stoma; UE, upper epidermis; X, xylem; XF, xylem fibers. PAS/toluidine blue. × 260.**

70 **Transverse section of a leaf of spike grass (*Distichlis* spp.). Contrast the distribution of these tissues with that shown in Figure 69. Each vascular bundle is surrounded by a well-defined sheath (BS), external to which lies the chlorenchyma (asterisks). Ribs of sclerenchyma (Sc) lie above and below each bundle and the upper epidermis contains bulliform cells (BC) whose turgor changes roll and unroll the leaf blade. This species grows in salt marshes and the leaves have salt glands (G). Ph, phloem; S, stoma; X, xylem. PAS/toluidine blue. × 260.**

areas of contact between the "outer" surfaces of pouches and/or stroma lamellae. As one might expect, the grana visible in the living cell correspond to those areas of the membrane system in which extensive contact between the pouches and stroma lamellae has formed large stacks of membrane (Figure 74, asterisks). If this interpretation is correct, one would expect to see occasional profiles through the regions in which pouches have originated from stroma lamellae, and careful examina-

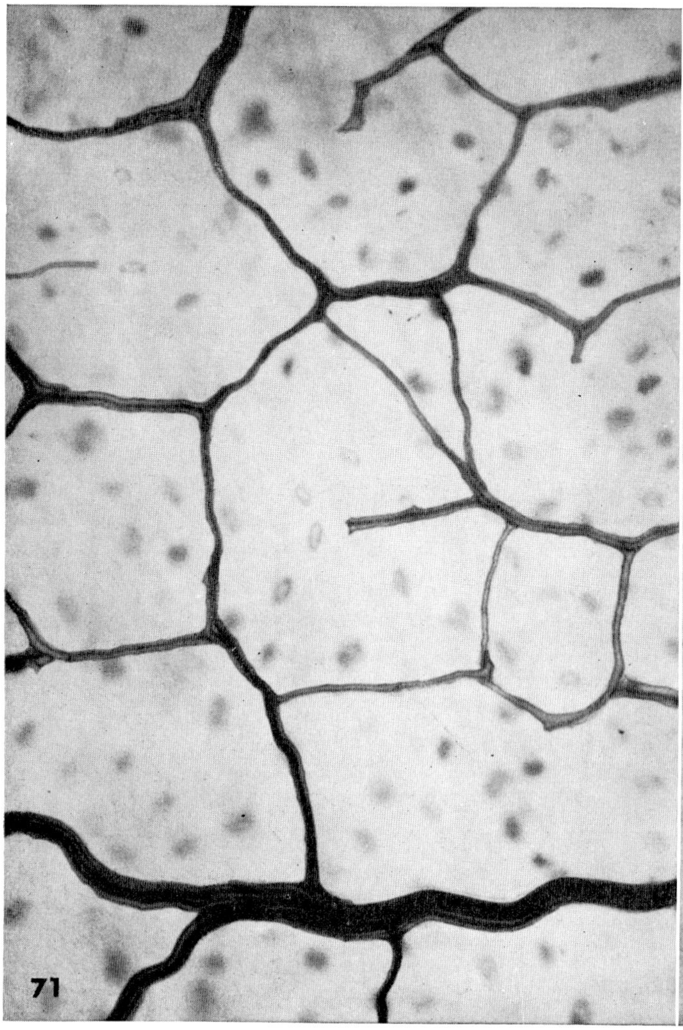

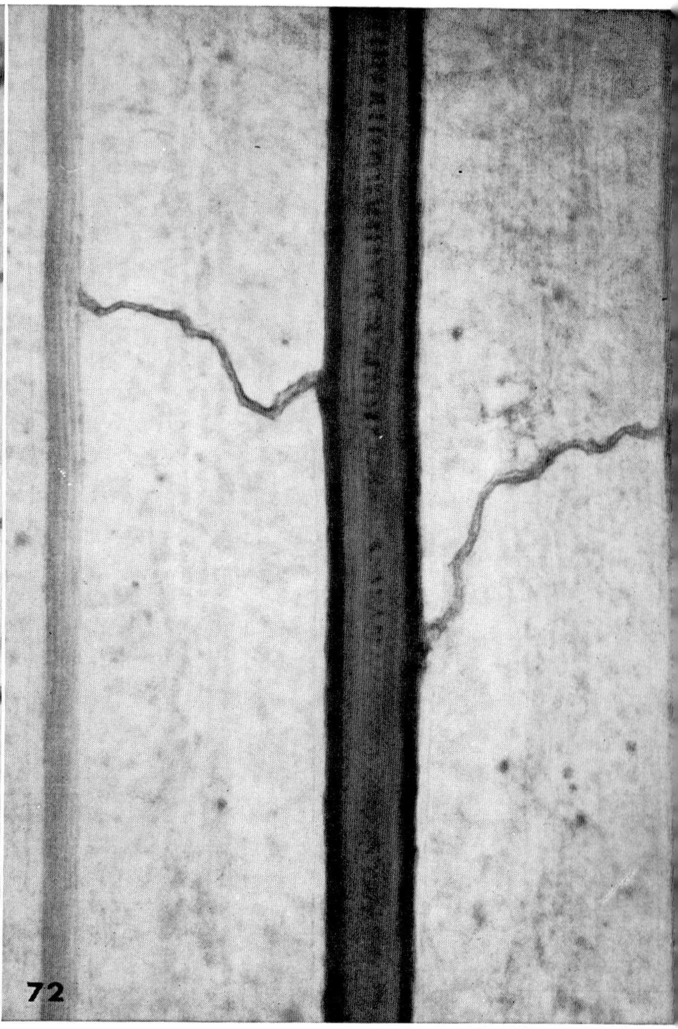

71 and **72** The arrangement of the vascular bundles in a leaf of snapdragon (*Antirrhinum majus*) (Figure 71) and oat (*Avena sativa*) (Figure 72), seen in surface view in cleared whole-mounts. Contrast the reticulate venation and occasional blind-endings of the dicotyledonous leaf (Figure 71) with the parallel venation (and small cross bridges) of the grass leaf (Figure 72). Compare with Figures 69 and 70. Tracheary elements stained with basic fuchsin. The stained structures which lie between the veins in Figure 71 are the guard cells of the stomata, which are somewhat out of focus. × 170.

73 Chloroplasts (P) in chlorenchyma of pea leaf. Note the large starch grains within the chloroplast (asterisks). IS, intercellular space; V, vacuole. PAS/toluidine blue. × 1400.

74 Electron micrograph of a chloroplast of a leaf of spinach (*Spinacia oleracea*). The chloroplast is surrounded by a double membrane (PM) and the internal membrane system is differentiated into grana (asterisks) and stroma lamellae (open arrows). Osmiophilic droplets (small black arrows) occur in the plastid stroma. The structure of the grana is shown in more detail in the inset as are regions of continuity between the grana and stroma lamellae (large solid arrows). CM, cell membrane; CW, cell wall; SG, starch grain; VM, vacuolar membrane. × 43,000; inset, × 68,500. (Courtesy Mr. A. D. Greenwood.)

74

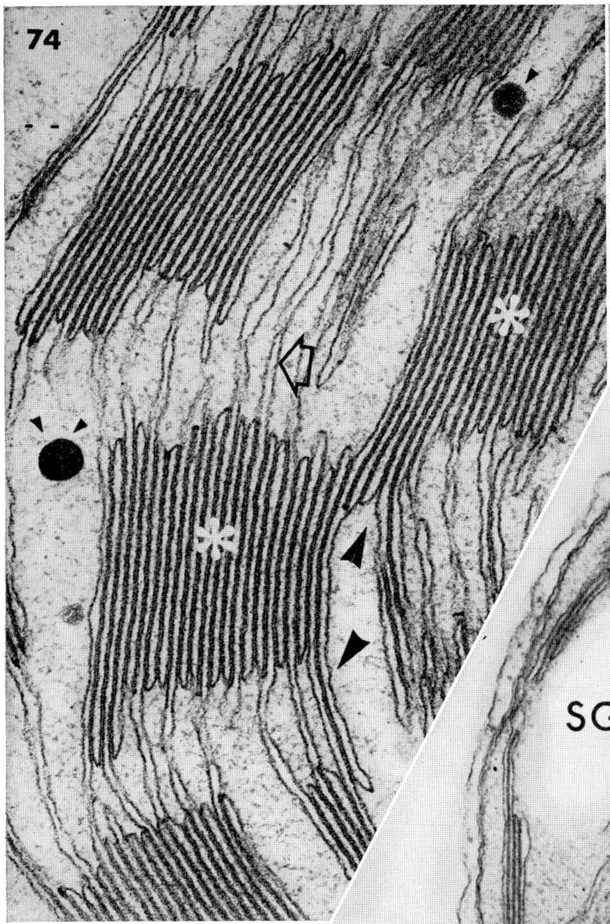

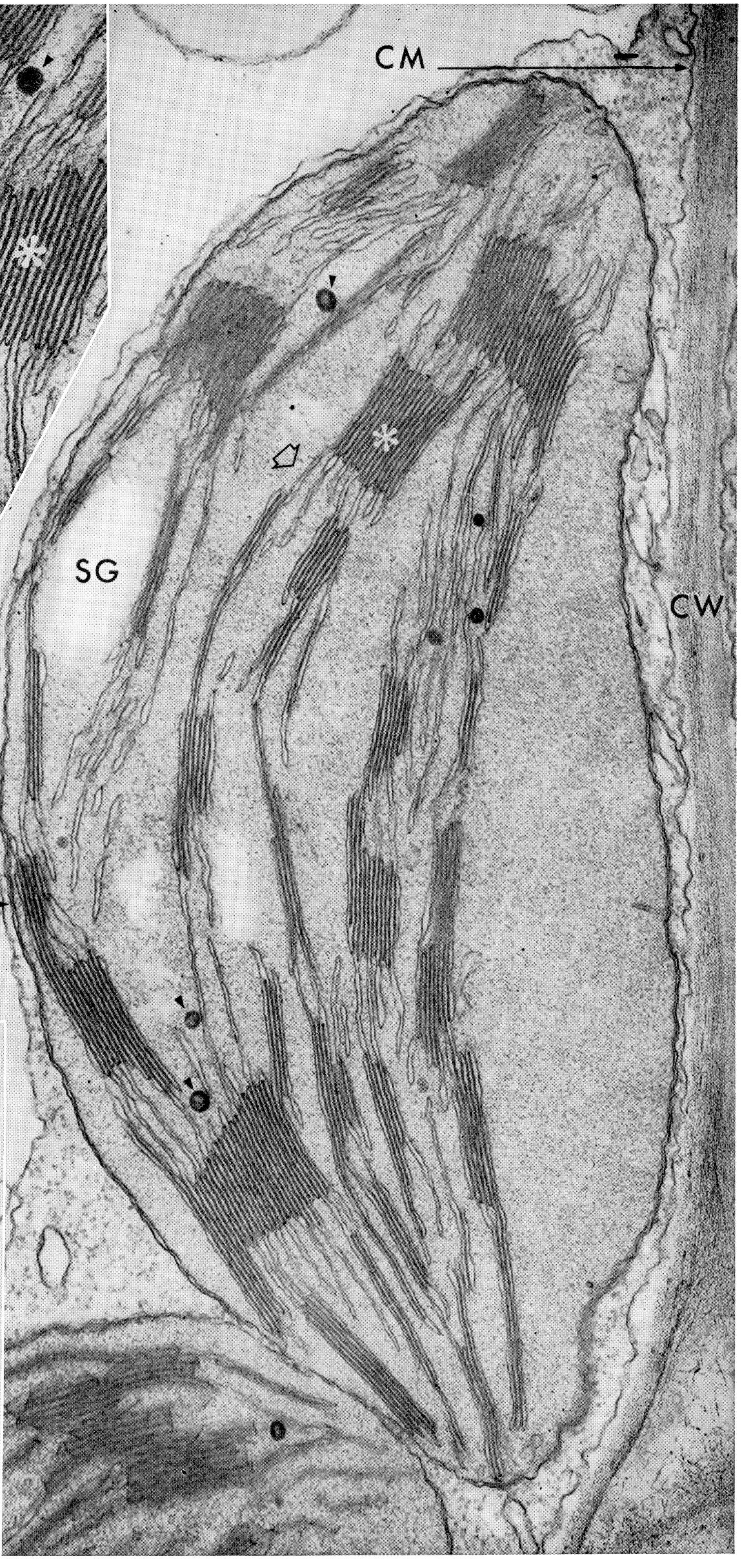

73

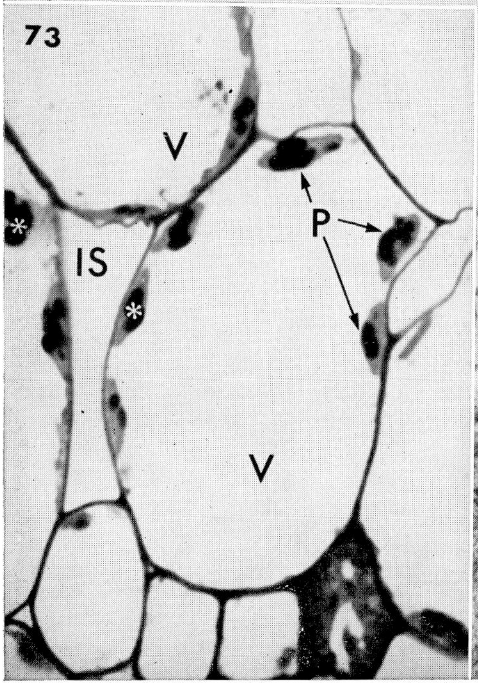

tion does reveal images which can be interpreted in that way (Figure 74, inset, arrows).

It is important to realize that the terms *granum* and *stroma lamella* have descriptive convenience only, and are not meant to imply that the membranes of these structures serve different functions, for such information is simply not available. If one traces the course of the stroma lamellae in Figure 74, it is clear that any one lamella may be connected to several grana, and it is possible, though difficult to establish with certainty, that the whole of the internal membrane system is interconnected. If this is the case, it follows that the space enclosed within the internal membrane system forms a continuum.

Elegant physiological experiments on intact leaves have established that the smallest unit of function in photosynthesis must be an aggregate that includes 200–300 chlorophyll molecules. In recent times, techniques have been developed which show that isolated chloroplasts contain all of the pigments and enzymes necessary to carry out the light and dark reactions of photosynthesis. Furthermore, analysis of fractions obtained from disrupted chloroplasts has shown that the membrane fraction contains the pigments used to trap light, as well as cytochromes and other enzyme systems that allow the energy made available from light absorption to be released in graded steps by electron transport. Just as in the mitochondrion, the enzymes of the membrane fraction carry out phosphorylation in conjunction with electron transport, and ATP is produced. The process is usually called *photophosphorylation* to distinguish it from oxidative phosphorylation in mitochondria. The enzyme that fixes carbon dioxide, and those that carry out the remaining complex interchanges of the Calvin cycle (a series of enzyme-mediated exchanges among sugar molecules) are located chiefly in the plastid stroma. In recent years, a small particle, about 180Å by 160Å, and 100Å thick, the *quantasome* (*5.4*), has been identified in the membranes isolated from chloroplasts. Biochemical analysis of these particles suggested initially that they might be the morphological expression of the smallest photosynthetic unit mentioned above, but this view has been challenged. The role of the *osmiophilic droplets* (Figure 74) remains obscure, though there is evidence that they are rich in lipids and quinones (*5.5*).

The basophilia of the chloroplast stroma is due in part to its content of RNA. In certain chloroplasts, the stroma is rich in small granules that resemble cytoplasmic ribosomes in appearance (these are not very evident in Figure 74, but the cytoplasmic ribosomes of this preparation are also weakly stained). These granules, somewhat smaller than cytoplasmic ribosomes, have been called *chloroplast ribosomes,* and have been shown to be rich in RNA (*5.6*). Chloroplast ribosomes are conspicuous in some plastids (*5.7*), and it has been shown that they can assume polyribosome configurations and can associate with the developing internal membrane system during greening of etiolated leaves (*5.8*). Chloroplasts also contain DNA, and they can incorporate suitable labelled precursors into RNA and protein (*5.9*).

Replication of plastids is usually confined in higher plants to division of the relatively simple plastids found in meristems. However, in some tissues, especially those of algae (*5.10*) and fern gametophytes, and during greening in etiolated leaves of higher plants, plastids of quite complex internal structure divide by fission. These facts, coupled with genetic evidence drawn from the inheritance of "plastid phenotypes" and the evidence

75 **Transverse section of outer epidermal cells of the coleoptile of wheat (*Triticum aestivum*), showing the location of the outer wall and pectin lamella (PL). N, nucleus; P, plastid; V, vacuole. Toluidine blue. × 1400.**

76 **Transverse section of outer epidermal cells of an oat coleoptile, fixed in glutaraldehyde/osmium tetroxide and stained with a cationic dye at high pH. This procedure stains the pectin lamella (PL) very strongly and also stains the cuticle (C). × 1950.**

77 **Electron micrograph of a section adjacent to that shown in Figure 76; the area outlined in the rectangle is seen in this figure. The structure of the pectin lamella, so evident in Figure 76, is obscured by the osmiophilia, and consequent electron densty, of the cuticle (C). The asterisks show cutin cystoliths in both Figures 76 and 77. P, plastid; V, vacuole. × 7000. (Figures 76 and 77 reproduced with permission from *Protoplasma*, 63, 1967, p. 393.)**

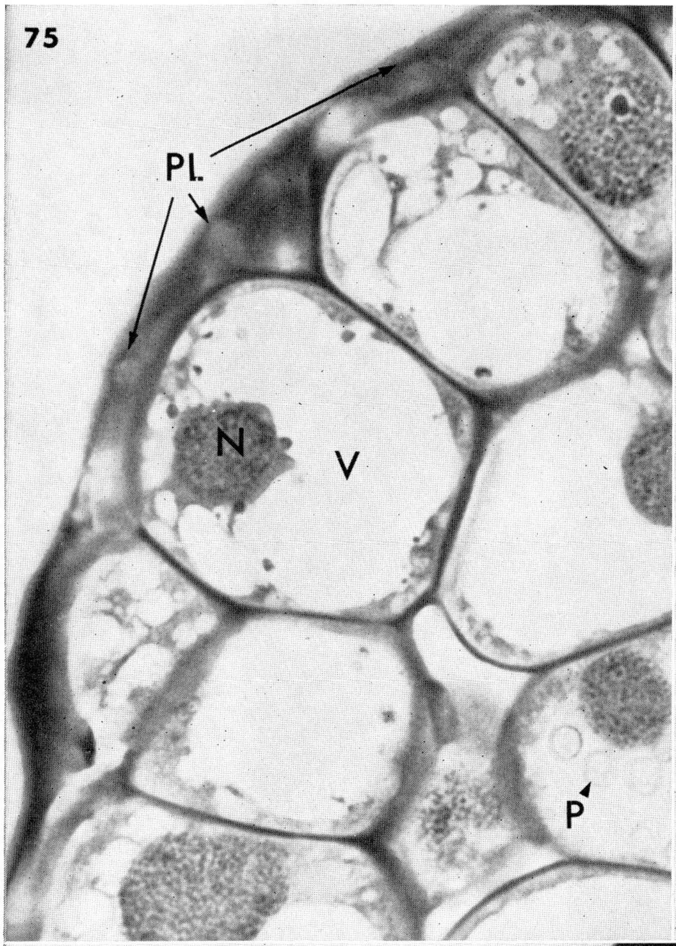
75
PL
N
V
P

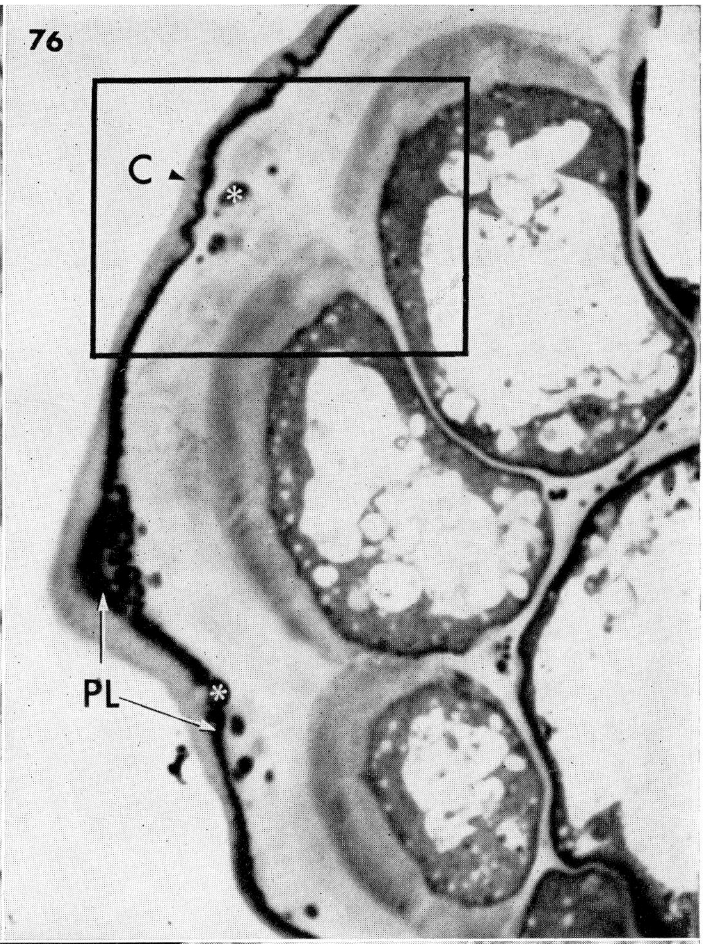
76
C
*
PL
*

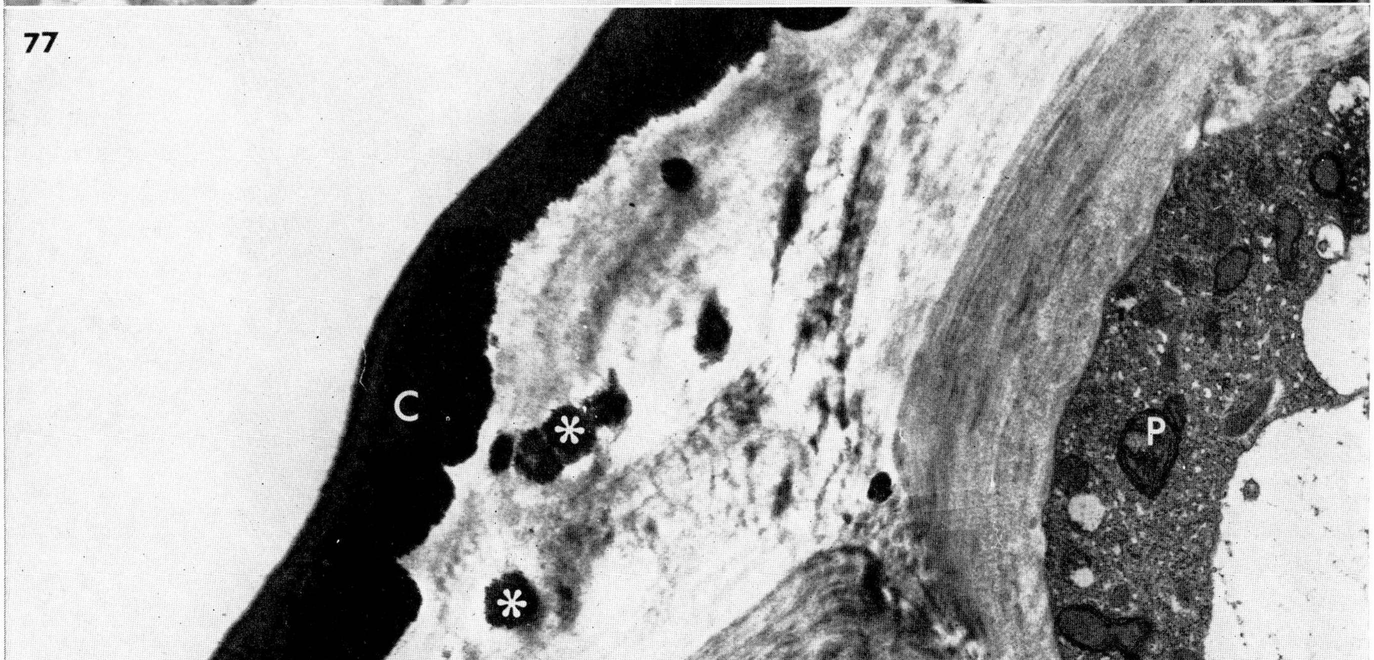
77
C
*
*
P
V

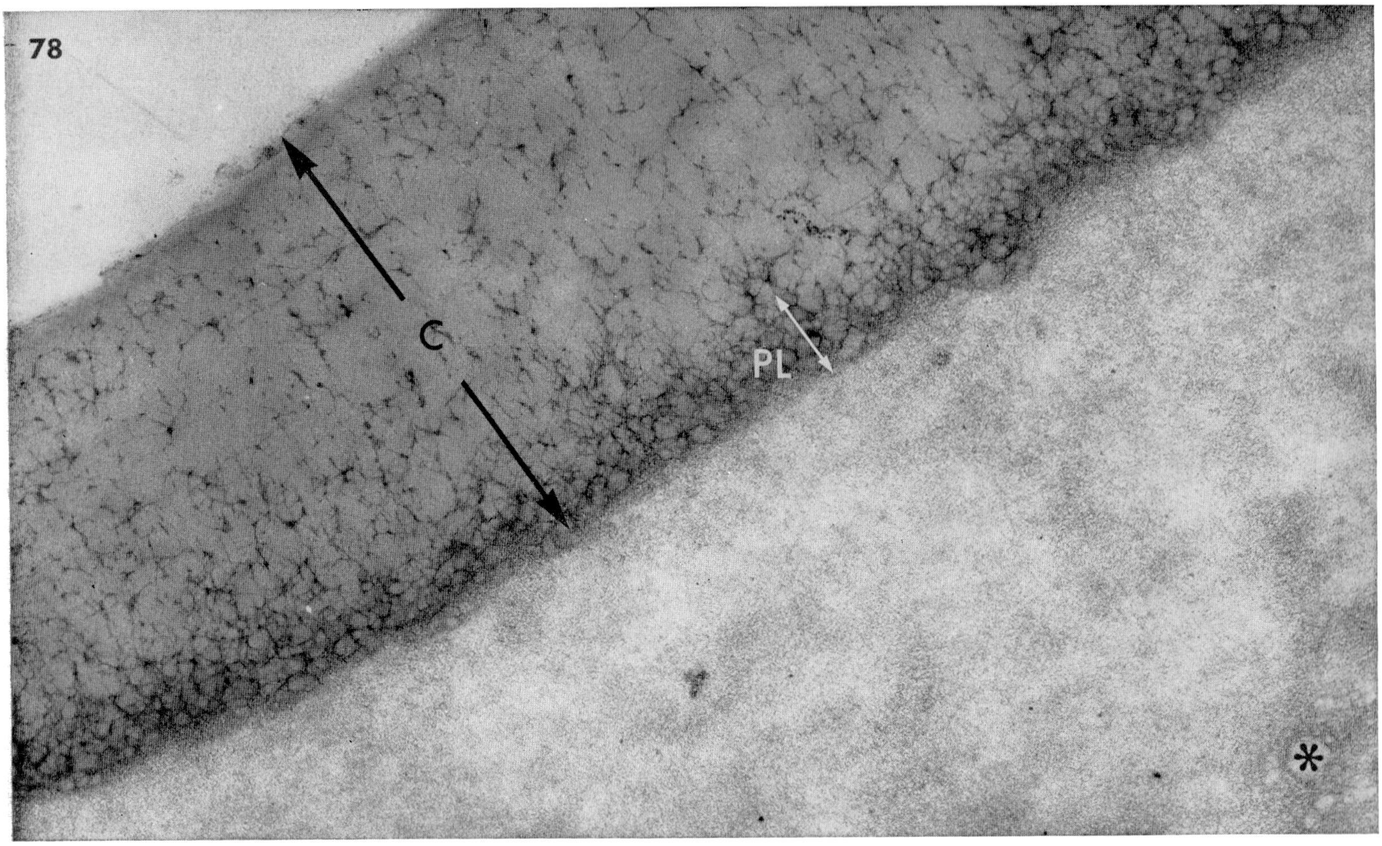

78 Electron micrograph of the cuticle (C) and pectin lamella (PL) in the outer epidermal cells of the coleoptile of wild oat (*Avena fatua*). The cuticle is less osmiophilic (and therefore less electron dense) and the pectin lamella is seen to consist of a reticulum of material (rich in polyuronides) that extends throughout the whole depth of the cuticle. A small cutin cystolith may be seen in the outer layer of the wall (asterisk). × 50,500. (Reproduced with permission from *Protoplasma*, 63, 1967, p. 395.)

of partial biochemical autonomy mentioned above, have led several investigators to speculate, as has been done for mitochondria, that plastids might also have evolved from an intracellular symbiont.

The Epidermis of the Leaf

Most leaves have an *upper* and a *lower epidermis* the cells of which may differ in size, type, and distribution (see Figures 69 and 70). Epidermal cells are usually highly vacuolated and longer than the adjacent mesophyll cells because as the leaf develops the rate of cell division declines first in the epidermal layers. (However, the cell divisions that produce *stomata* and *hairs* (see Figure 68) often occur quite late in leaf development.) Although the bulk of cell division ceases, DNA synthesis does not, and the large, chromatin-rich nuclei of epidermal cells (Figure 75) are often endopolyploid, with DNA levels up to thirty-two times the haploid amount.

79 The upper epidermis of a fresh pea leaf, seen in surface view. Note the stomata and the paired guard cells and the squamous outline of the anticlinal walls of the epidermal cells except where they overlie a vascular bundle (upper right corner). Toluidine blue. × 157.

80 Electron micrograph of a replica of the wax on the surface of a pea leaf. One may again detect the squamous outline of the walls of an epidermal cell and also the position of two stomata. × 2500. *Inset:* Detail of Figure 80 at high magnification to show the structure of the wax. Inset, × 40,000. (Courtesy Dr. B. E. Juniper.)

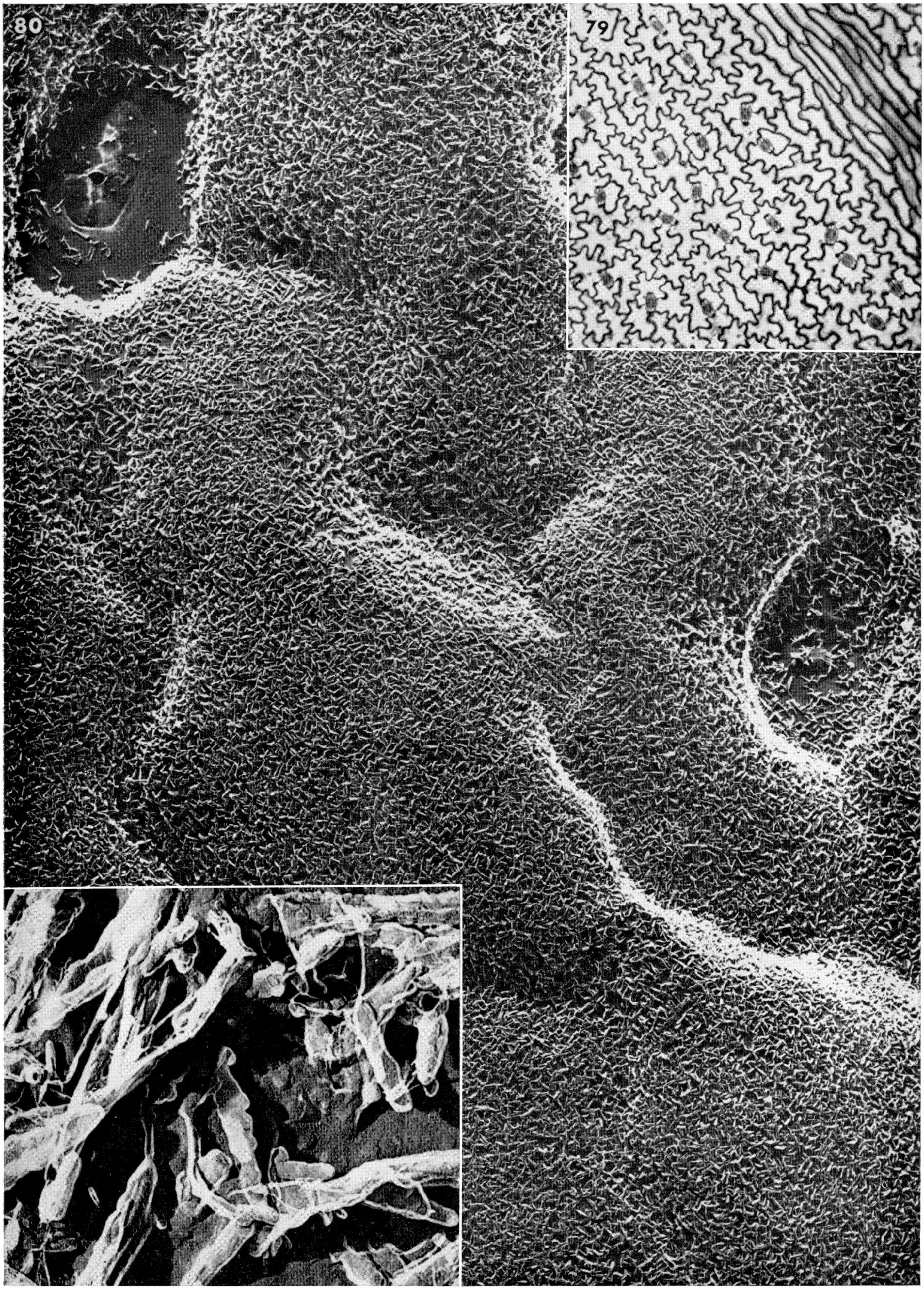

80

79

Although epidermal cells carry out a variety of functions, we wish to concentrate here upon their protective properties. Unfortunately, the same features (thick walls and lipid-impregnated cuticles) that allow epidermal cells to resist desiccation and invasion by pathogens also make them difficult specimens for electron microscopy. The illustrations of wall and cuticle structure which follow are from oat coleoptiles whose epidermal cells proved to be simpler to prepare for study than those of leaves. Although the coleoptile is not usually a photosynthetic organ, the cells of its outer epidermis carry out similar functions to those of the epidermal cells of leaves, and they have a similar wall structure.

The outer wall of epidermal cells is invariably the thickest, and all of the walls stain strongly with cationic dyes (Figure 75). Water loss through the surface of epidermal cells is restricted by a *cuticle* (*5.11*), a layer impregnated with lipids of high molecular weight. An accumulation of polyuronides termed the *pectin lamella* by early histologists marks the boundary of the cuticle and the wall proper (Figure 75). Both the cuticle and the pectin lamella are more obvious in thin (0.25μ thick) sections from tissues treated with OsO_4 and stained with a cationic dye at high pH (Figure 76). The appearance in the electron microscope of the area outlined by the rectangle in Figure 76 is shown in an adjacent section in Figure 77. The intense osmiophilia and consequent electron density of the cuticular lipids obscures detail at low magnification, and the less osmiophilic cuticle of a different species of oat is shown at higher magnification in Figure 78. This shows that the "pectin lamella" is but the basal part of a reticulum, rich in polyuronides, which extends right through the cuticle to its outer surface (*5.12*). It is most likely that this reticulum provides the pathway by which water is lost as *cuticular transpiration* and that along the same route, water soluble substances can penetrate.

The anticlinal walls of the epidermal cells frequently produce a squamous pattern in surface view, except where they overlie a vascular bundle (Figure 79). The surface of the cuticle is covered with wax in many species; Figure 80 is an electron micrograph of a replica of the wax structure on a pea leaf. The outline of individual epidermal cells and the depressions that mark the position of the stomata are evident. At higher magnification, the individual rodlets and plates of wax may be resolved (Figure 80, inset). The patterns observed in the leaf wax are reasonably constant within any one species and for a particular part of the plant, but the pattern varies sharply among species and is also sensitive to the environment (temperature, wind, and rate of water loss) (*5.13*). The wax coating certainly makes the leaf less wettable and probably renders it more resistant to invasion by pathogens. It is remarkable that the complex lipids of the cuticular framework and the wax can migrate through the hydrophilic inner layers of the wall to their final location. What role the polyuronide-rich reticulum plays in this process is still to be elucidated.

Many plants produce succulent fruits that help to ensure that the seeds of these plants are more widely dispersed by the animals that eat them. In many cases, the succulent tissues are protected from desiccation by a layer of epidermal cells whose outer wall may become very thick (Figure 83). In addition, this thick layer of polysaccharide-rich cuticle is often impregnated with wax, seen in replica in the electron microscope (Figures 81 and 82).

The Stomata

The epidermal cells restrict the exchange of all gases, including carbon dioxide, between the internal air spaces and the external environment. If the whole of the leaf surface were covered, photosynthesis would be limited by the amount of carbon dioxide that could diffuse through the surface. The stomata afford a partial solution to this problem and furthermore, they allow the exchange

81 and 82 **The surface wax of an apple fruit seen as a replica in the electron microscope. Natural surface (Figure 81) and the same surface after polishing (Figure 82). Figure 81, × 11,500, Figure 82 × 18,000. (Courtesy Dr. D. S. Skene. Reproduced with permission from *Annals of Botany*, N.S., 27, 1963, pp. 581–587.)**

83 **Section of the epidermis of an apple fruit showing the thick layer of extracellular material (ECM) which overlies the epidermal cells. Toluidine blue. × 770.**

81

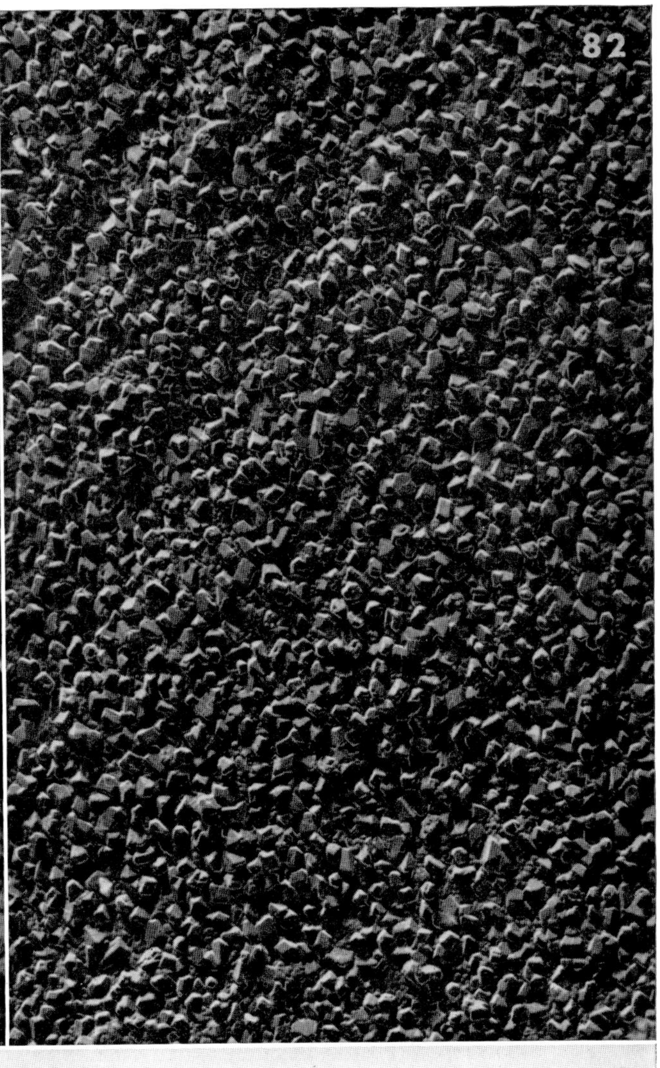
82

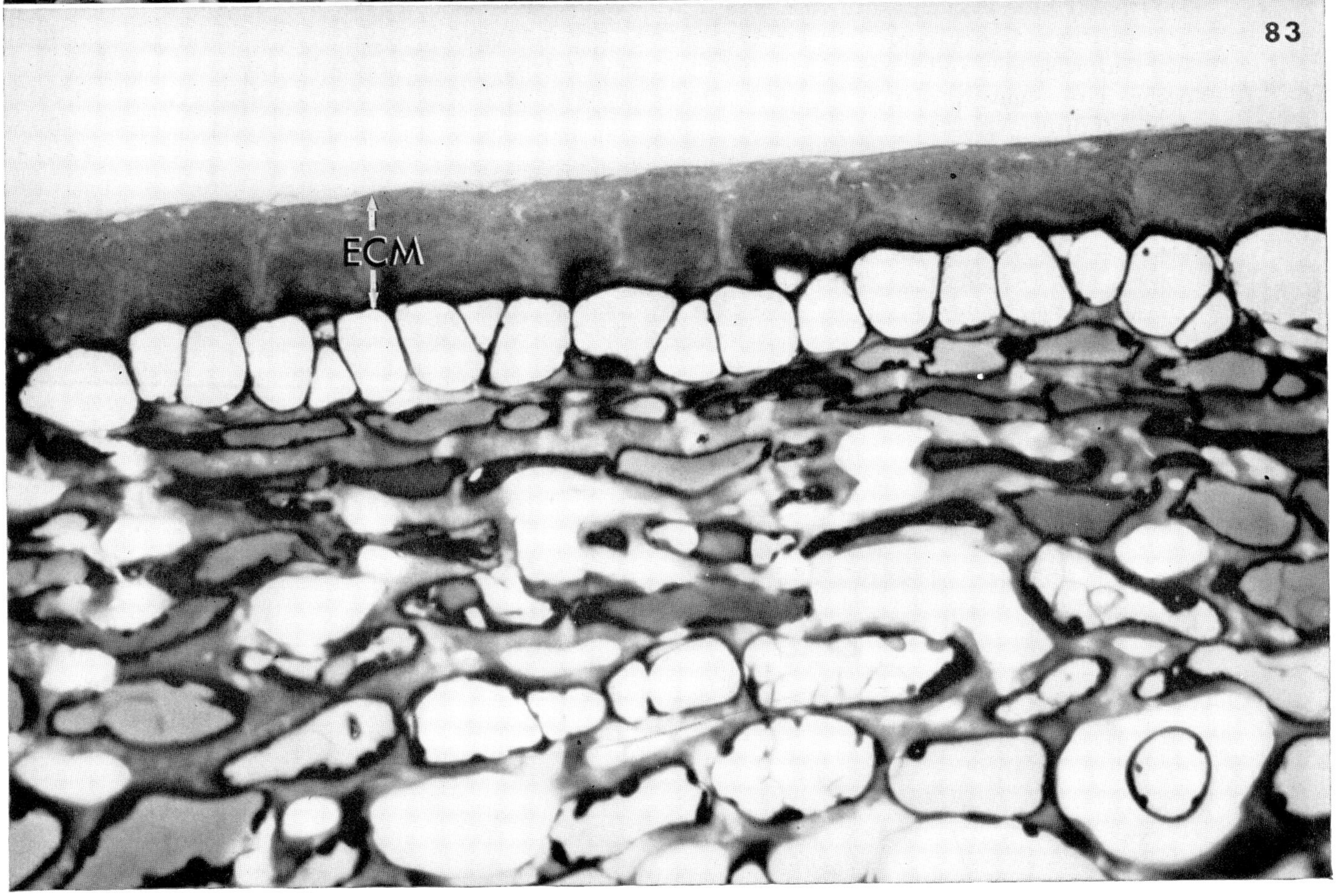
83
ECM

process to be regulated by the plant. Each stoma consists of a pore surrounded by two *guard cells.* The aperture of this pore is regulated by the turgor of the guard cells. This is possible only because the guard cell walls are asymmetrically thickened with the thicker part of the walls adjacent to the pore. The thicker parts of the wall are less extensible than the thinner parts, and the guard cells change their shape as their turgor varies, opening or closing the pore between them. The guard cells respond to a number of stimuli, including water stress, carbon dioxide concentration and mechanical shock. In some plants stomatal aperture varies on a rhythmic basis (*5.14*).

In the electron microscope, guard cells are seen to have a large number of mitochondria and a well-developed endoplasmic reticulum (Figure 84). Dictyosomes are also numerous in Figure 84, but these cells were from a young bean leaf and the guard cell walls were still being thickened. No completely satisfactory explanation for the mechanism that regulates the turgor of the guard cells has yet been advanced. Since mitochondria are known to be very effective in regulating solute concentration in certain cell types, perhaps the large numbers of mitochondria in the guard cells control turgor by a similar process.

The number of stomata per unit area of leaf varies with the species and is often greater in the lower epidermis than in the upper. The distribution of stomata in Figure 79 is characteristic of many dicotyledons and the pattern in which they occur is precisely regulated during the development of all leaves. In many species, the outer surface of the guard cells is adorned by pegs or ridges (Figure 84, inset). It was predicted on theoretical grounds and then conclusively demonstrated by experiments with aqueous solutions of fluorescent dyes that these elaborations of the cuticle prevent rain water from entering the aperture (*5.15*).

Other Specialized Epidermal Structures: Glands and Hairs

Glands and hairs (*5.16*) are found in the epidermis of many species. They range in complexity from the stinging "hairs" of nettles (Figure 68), to glands that may secrete salts, sugars, oils, resins, or mucilages. The function of many glands and most hairs is still unknown.

The glands on the leaf of *Odontites verna* (Figures 85–88) have been studied recently in some detail (*5.17*). This plant is a root hemiparasite, and forms a connection to the xylem of the host through a haustorial pad of tissue. Through this connection it diverts a fraction of the host's transpiration stream. If the host xylem is labelled with a fluorescent dye, the dye soon appears in the exudate from the leaf glands of the hemiparasite. Comparison of the solutes present in the host xylem, parasite xylem, and in the gland exudate suggests that the glands are very effective in removing materials from the transpiration stream. These glands help to ensure that the parasite can compete with the host for the host's transpiration stream because they secrete fluid in excess of that lost by evaporation from the other epidermal cells of the parasite leaf. In addition, it is likely that the glands accumulate the last traces of organic compounds from the exudate and utilize these compounds for their own syntheses (see Figure 88).

The Abscission of Leaves

Leaves are short-lived organs, rarely functioning for more than a few years even in "evergreen" species. In deciduous plants the life-span of leaves is only one growing season and the shedding of leaves in the autumn is a spectacular phenomenon of temperate regions.

The remarkable feature about leaf fall is that it is not a haphazard process but rather appears to result from a highly correlated hormone-mediated series of events within the plant (*5.18, 5.19, 5.20*). Even the point of breakage between the leaf and the stem is not located randomly but occurs in a specific abscission zone. One of these zones is present across the petiole near its connection to the node. Compound leaves frequently have additional breaking points at the bases of the leaflets. In many species the areas where abscission will eventually occur can be distinguished in the petioles of young leaves before the event takes place. A potential abscission zone is usually characterized externally by a small constriction on one side of the petiole (Figure 89) and internally by the presence of a thin disc of small cells running across the petiole cortex (Figure 90).

Leaf abscission is always preceded by leaf senescence, a complex event characterized by the break-

84 Electron micrograph of a transverse section through part of a pair of guard cells in the leaf of *Phaseolus vulgaris*. The cells have an abundance of endoplasmic reticulum (ER), dictyosomes (D), and mitochondria (which have not been well preserved; M*). N, nucleus. The inset shows the appearance in the light microscope of a comparable section of guard cells from a pea leaf. Note the cuticular projections (CPr) in both figures. × 27,600. Inset, IS, intercellular space. Toluidine blue. × 1820.

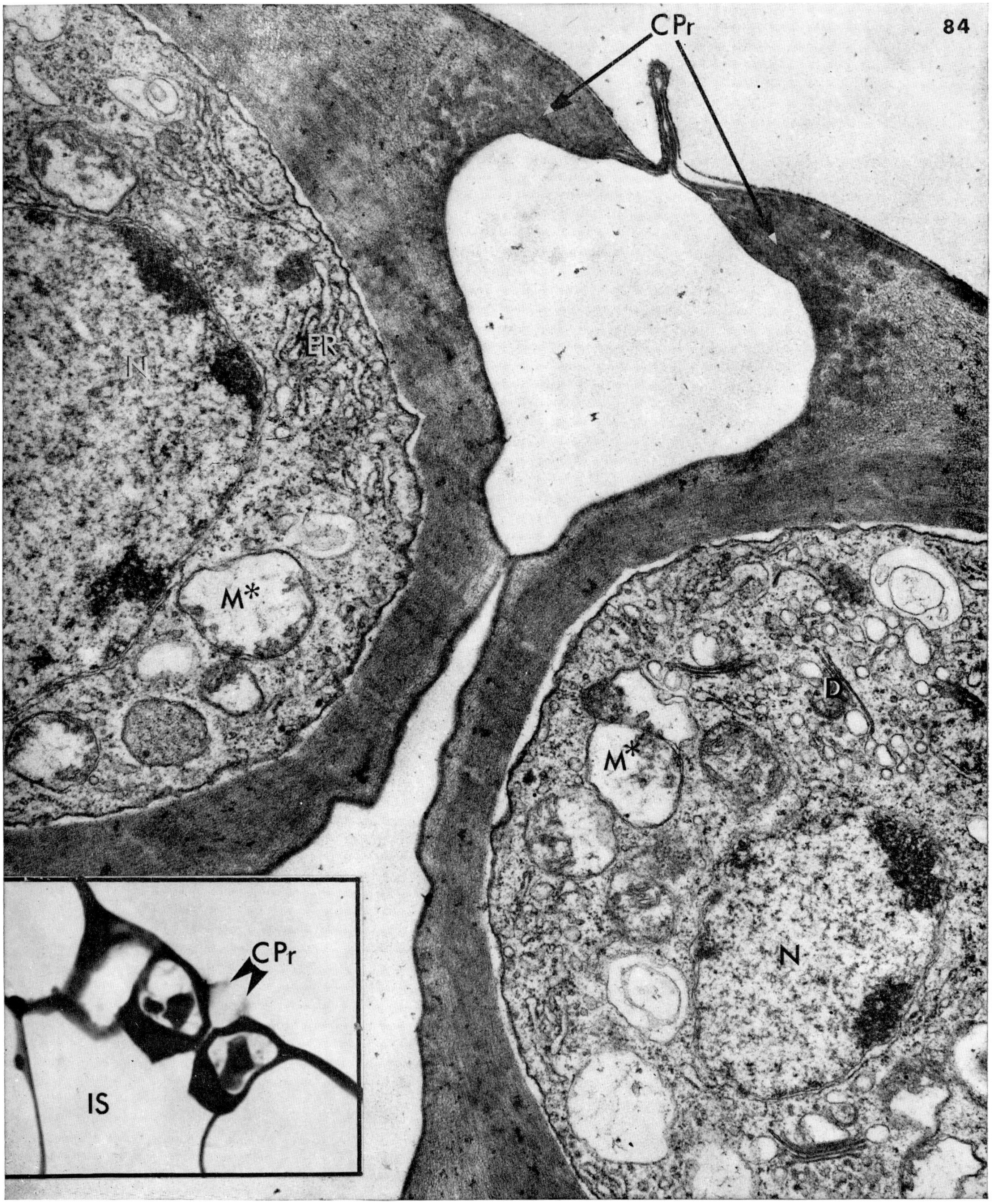

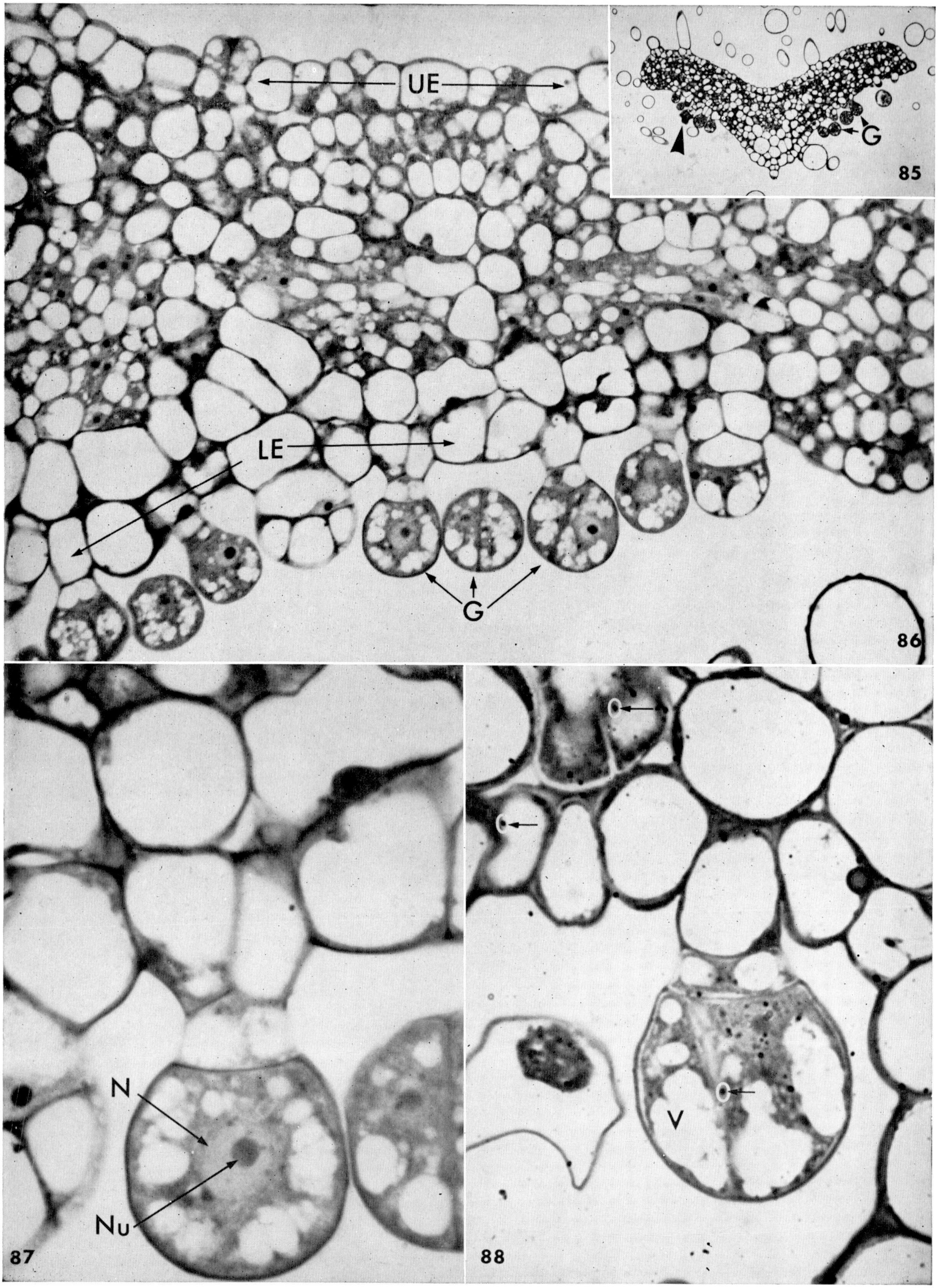
UE
G
85
LE
G
86
N
Nu
87
V
88

89 Stages of petiole abscission in isolated nodes of *Coleus*. In the segments from right to left, the leaf blade had been removed for 53, 78, and 96 hours respectively. The abscission zone is indicated by the arrows. × 2.3. (Preparation courtesy Mr. R. D. Muir.)

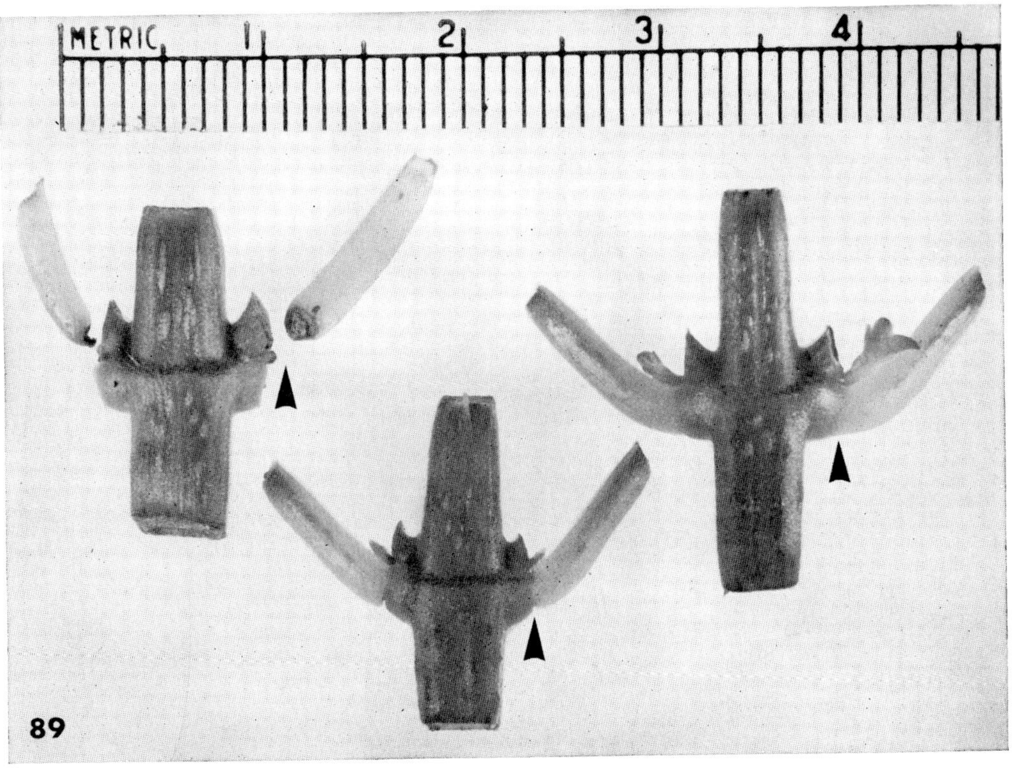

down of the chloroplasts and the disappearance of chlorophyll, a reduction in protein content, and not infrequently, by an increase in the production of anthocyanin pigments.

Removal of the leaf blade results in the abscission of even young petioles. Since this phenomenon will occur *in vitro* in "explants" of a node and its attached petioles (Figure 89), such *in vitro* systems have been widely used in investigations of abscission in different plants. A few hours after its leaf blade has been removed a petiole begins to turn yellow. By the time the petiole falls, the chlorophyll has disappeared. However, chlorosis does not occur proximal (toward the stem) to the abscission zone, and the break between yellow and green tissue sharply defines this region. In *Coleus* (Figure 91), as in most other species, abscission is preceded by stimulation of cell division in the portion of the abscission zone that will remain with the stem.

Actual abscission, which begins in *Coleus* explants about sixty-five hours after leaf removal, is first apparent as a parting of adjacent cell walls,

85-88 Glandular hairs on the leaf of the root-hemiparasite *Odontites verna*. All stained with toluidine blue. (Section courtesy Dr. R. N. Govier, Dr. J. S. Pate, and Mr. J. Brown.)

85 Transverse section of young leaf showing the distribution of several glands (G). The glands are ephemeral and wither within a few days (arrow). × 90.

86 An array of glands on the lower epidermis (LE) of the leaf. UE, upper epidermis. × 740.

87 The structure of one gland at high magnification. The glands secrete a fluid derived ultimately from the transpiration stream of the host plant and are believed to assist the parasite in diverting the host's transpiration stream into the parasite tissues. N, nucleus; Nu, nucleolus. × 2000.

88 An autoradiogram that shows that a metabolite (^{3}H-leucine) fed via the transpiration stream, is indeed incorporated into insoluble material by the tissues of the young leaf and by the glands (some of the numerous silver grains are circled at arrows). V, vacuoles. × 2000.

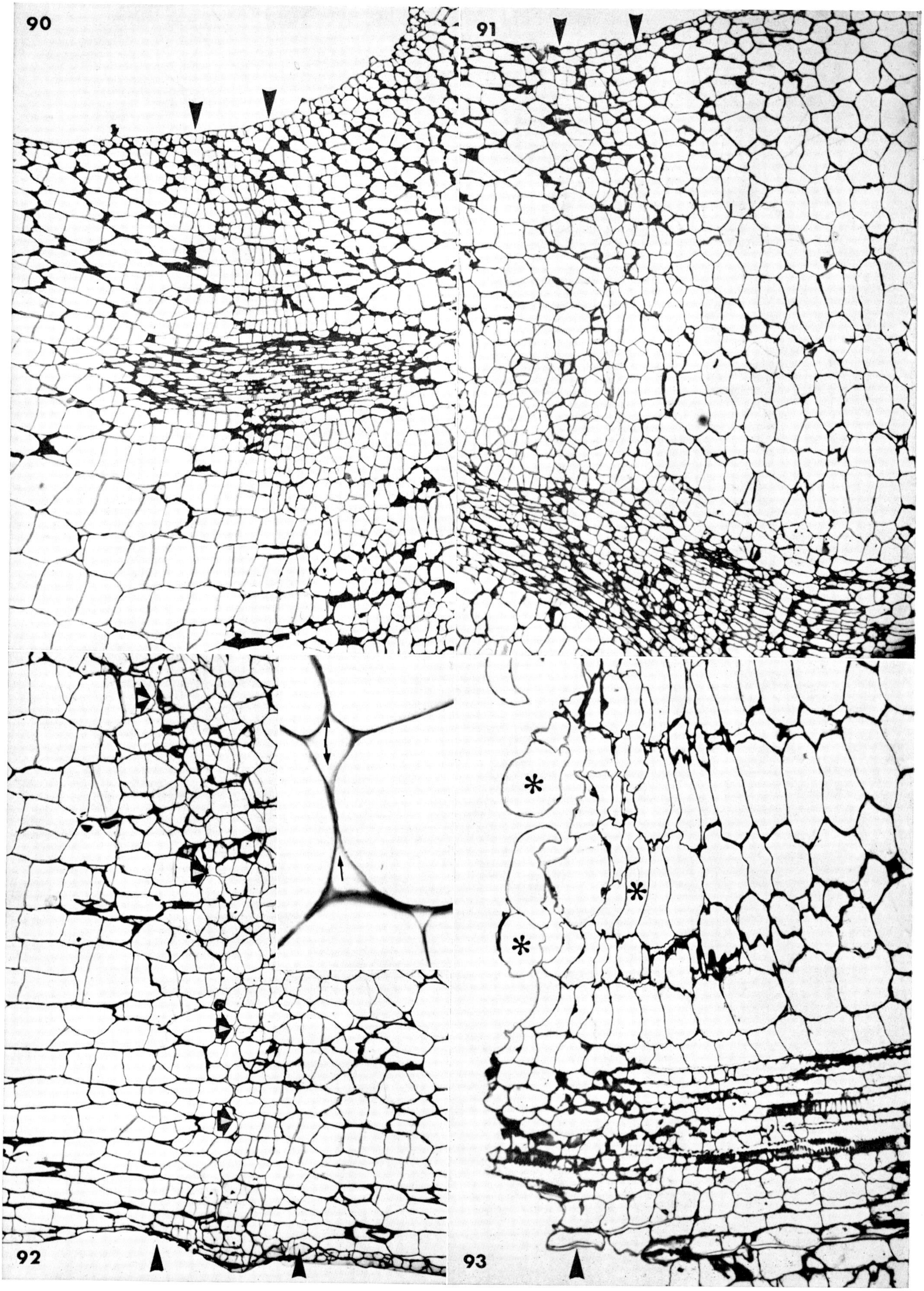

90
91
92
93

90-93 **Longitudinal sections of *Coleus* petioles showing a portion of the abscission zone (Figure 89, arrows) at different stages of induced abscission *in vivo*. In each case the thickness of the abscission zone is indicated by black arrows and the "blade" end of the petiole is toward the left of the picture. Toluidine blue. (Sections courtesy Mr. R. D. Muir.)**

90 **Normal petiole with intact leaf blade showing potential abscission zone. × 75.**

91 **Leaf blade removed for 53 hours. × 75.**

92 **72 hours after the removal of the leaf blade, the cell walls are separating (black and white arrows) along a cleavage plane through the abscission zone. Inset shows details of separation. × 75. Inset, × 1050.**

93 **Stump remaining just after the petiole has abscised, 73 hours after removal of the leaf blade. Cleavage has been *between* the walls of the cortical cells, and intact cells (asterisks) remain on the surface at the "stem" end of the petiole. × 100.**

approximately along a single plane across the cortex (Figure 92 and inset). Only the walls separate and the cortical cells remain intact (Figure 93). When this separation is complete in the cortex, the weight of the petiole appears to break the vascular bundle mechanically and the petiole falls off. It has been found that even a small portion of healthy leaf blade left at the end of the petiole will greatly delay or prevent abscission. This retarding effect can also be obtained by the application of auxin to a debladed petiole.

Recent work has shown that the application of cytokinins to a debladed petiole will also delay abscission, whereas two other hormones—gibberellin and abscisic acid—increase the rate of petiole abscission when applied in the same way. Oddly enough, abscisic acid, which is a stimulator of abscission in many plants, can be extracted from *both* senescent and healthy leaves. It seems, therefore, that the action of healthy leaves in preventing their own abscission may not be so much the production of an abscission-retarding hormone per se as the maintenance of a hormone balance in the petiole which favours inhibition of abscission.

Whatever the nature of the hormonal control mechanism, the immediate cause of abscission is a localized enzymatic breakdown of portions of the adjoining cell walls on either side of the abscission zone, allowing separation of the cells.

GENERAL REFERENCES

Esau, K. *Plant Anatomy*, Chap. 16. New York: John Wiley & Sons, 1965.

Foster, A. S. *Practical Plant Anatomy*, Chap. 14. Princeton, N. J.: D. van Nostrand Co., 1949.

Kirk, J. T. O., and R. A. E. Tilney-Bassett. *The Plastids*. London: W. H. Freeman and Co., 1967.

Milthorpe, F. L. *The Growth of Leaves*. London: Thornton Butterworth, Ltd., 1956.

REVIEWS AND RESEARCH PAPERS

5.1 Sharman, B. C. "Leaf and bud initiation in the Gramineae." *Bot. Gaz.*, 106 (1945), 269–289.

5.2 Sifton, H. B. "Air-space tissue in plants," *Bot. Rev.*, 11 (1945), 108–143.

5.3 WEHRMEYER, W. "Über Membranbildungsprozesse im Chloroplasten. II. Zur Entstehung der Grana durch Membranüberschiedung." *Planta,* 63 (1964), 13–30.

5.4 PARK, R. B. "The chloroplast," in *Plant Biochemistry,* ed. J. Bonner and J. E. Varner. New York: Academic Press, 1965.

5.5 LICHTENTHALER, H. K., AND B. SPREY. "Über die osmiophilen globulären Lipideinschlüsse der Chloroplasten." *Zeit. für Naturforsch.,* 21b (1966), 690–697.

5.6 JACOBSEN, A. B., H. SWIFT, AND L. BOGORAD. "Cytochemical studies concerning the occurrence and distribution of RNA in plastids of *Zea Mays.*" *J. Cell Biol.,* 17 (1963), 557–570.

5.7 GUNNING, B. E. S. "The fine structure of chloroplast stroma following aldehyde osmium-tetroxide fixation." *J. Cell Biol.,* 24 (1965), 79–93.

5.8 BROWN, F. A. M., AND B. E. S. GUNNING. "Distribution of Ribosome-like particles in *Avena* plastids." In *Symposium on Biochemistry of Chloroplasts* (Aberystwyth, 1965), ed. T. W. Goodwin. London: Academic Press, 1967.

5.9 SWIFT, H. "Nucleic acids of mitochondria and chloroplasts," *Amer. Nat.,* 99 (1965), 201–227.

5.10 GREEN, P. B. "Cinematic observation on the growth and division of chloroplasts in *Nitella,*" *Amer. J. Bot.* 51 (1964), 334–342.

5.11 ROELOFSEN, P. A. "On the submicroscopic structure of cuticular cell walls," *Acta Botan. Neerl.* 1 (1952), 99–114.

5.12 O'BRIEN, T. P. "Observations on the fine structure of the oat coleoptile. I. The epidermal cells of the extreme apex," *Protoplasma,* 63 (1967), 385–416.

5.13 JUNIPER, B. E. "The surfaces of plants," *Endeavour,* 18 (1959), 20–25.

5.14 HEATH, O. V. S. "The water relations of stomatal cells and the mechanisms of stomatal movement," in *Plant Physiology,* II. ed. F. C. Steward. New York: Academic Press, 1959.

5.15 CURRIER, H. B., AND C. D. DYBING. "Foliar penetration of herbicides: A review and present status," *Weeds,* 7 (1959), 195–213.

5.16 UPHOF, J. C. T., AND K. HUMMEL. "Plant hairs," in *Encyclopaedia of Plant Anatomy,* IV, Pt. 5. Berlin-Nikolassee: Gebrüder-Borntrager, 1962.

5.17 GOVIER, R. N., M. D. NELSON, AND J. S. PATE. "Hemiparastic nutrition in angiosperms," *New Phyt.,* 66 (1967), 285–297.

5.18 ADDICOTT, F. T. "Physiology of abscission," *Encyclopaedia of Plant Physiology,* XV, Pt. 2, ed. W. Ruhland. Berlin: Springer-Verlag, 1965.

5.19 BORNMAN, C. H., A. R. SPURR, AND F. T. ADDICOTT. "Abscission, auxin, and gibberellin effects on the developmental aspects of abscission in cotton (*Gossypium hirsutum*)," *Amer. J. Bot.,* 54 (1967), 125–135.

5.20 OSBORNE, D. J. "Regulatory mechanisms in senescence and abscission," in *Proceedings of the Sixth International Conference on Plant Growth Substances,* ed. F. Wightman and G. Setterfield. Ottawa: Runge Press (On press.)

94 Winter twig of sycamore (*Acer pseudoplatanus*). × 3

6
Buds

BUDS ARE embryonic shoots. They vary in maturity from those that consist of little more than an apical meristem to those that have developed many immature leaves and have differentiated nodes and short internodes. Buds differ from growing shoots, however, in having all their tissues in a state of arrested development.

Buds are most evident on winter twigs (Figure 94). Here both apical and lateral buds are obvious because of their protective envelopes of bud scales, or *cataphylls.* These cataphylls develop on the flanks of the apical meristem in the same manner as leaves (see Chapter 4) shortly before growth of the shoot-tip stops. The cataphylls are frequently thick and leathery in texture, and their epidermal cells are often specialized for the production and secretion of large amounts of extracellular material (Figure 95). This material waterproofs the buds and may help to prevent desiccation of the internal tissues and damage from the cold.

A characteristic common to all buds is their lack of activity, or dormancy. In the case of winter buds this dormancy is obviously an adaptation that ensures that the meristematic tissues of the shoot apices survive the rigours of the cold season. Seasonal dormancy (*4.11*) is usually under photoperiodic control, being induced by long nights. Once they are dormant, winter buds must often be subjected to cold treatments of varying degree and duration before they will respond to shorter nights and higher temperatures, break their dormancy, and grow. The dormancy of winter buds is broken dramatically in the spring as growth and differentiation resume in the enclosed embryonic tissues. The old bud disappears (see Figure 57) as it is transformed into a growing shoot which subsequently forms new buds.

Although formation and growth of buds is controlled by the environment, induction of dormancy and its release must be mediated by hormones (*6.1*). External applications of cytokinins or gibberellins to dormant buds have been successful in inducing bud-break in some cases but have failed in others. Recently it has been shown that the application of abscisic acid, a hormone also involved in leaf abscission, can induce a growing shoot to form a dormant bud (*6.1, 6.2*).

The removal of a vigorously growing shoot apex always results in the release from dormancy of one or more of the lateral buds. In most cases the dominating effect of the intact apex can be reimposed if auxin is applied promptly to the apical stump, and it appears that it is the auxin flowing from the actively growing apices that provides the dominating influence on the lateral buds. The action of auxin in this instance is controversial (*6.3, 6.4, 6.5, 6.6*), but it appears that the auxin stimulates the nodal tissue to produce ethylene, which is the ultimate inhibiting agent.

In the twig illustrated in Figure 94, the apical bud is noticeably larger than the lateral buds. This is the usual situation and illustrates the phenomenon of apical dominance, a puzzle that has intrigued physiologists and morphologists for many years. The shoot apex of almost all plants exerts some measure of inhibition on the development of lateral branches. In some cases the lateral branches may simply be shortened, but in extreme cases the main apex may completely inhibit the growth of lateral shoots so that they remain as dormant buds. In the case of the sycamore shoot (Figure 94), the inhibition is only partial, and the lateral buds will develop into short branches. Even in the winter twig, the inhibitory effect of the apex is reflected in the smaller size of the lateral buds.

Whereas bud dormancy imposed by environmental factors is usually short-lived and seasonal, that imposed by apical dominance may last for the lifetime of the apex. In some plants, many potential shoots are held as buds by a vigorous shoot apex and their latent morphology is expressed only if the dominant apex is damaged. Commonly, buds escape from the effects of the dominating apex as it grows further away from them, so that when they are more than a certain distance from the apex, they grow out into lateral branches. These lateral branches become progressively freer from the effects of the main apex as both the laterals and the main apex continue to grow. Thus the form of plants is very much a reflection of the degree of dominance that the growing apices retain over the buds and lateral branches. It is worthwhile noting that the dormancy induced by environmental conditions appears to be superimposed on the inherent apical dominance. The removal of a dominant bud from a woody shoot does not release the inhibited laterals until their seasonal dormancy has also been broken.

An interesting loss of apical dominance occurs in some shrubby plants such as lilac (*6.7*). In these plants an apical bud is formed which temporarily restricts the growth of the lateral buds below it. However, this terminal bud eventually abscises and

95 LS of a bud scale of lilac. ECM, extracellular material. PAS/toluidine blue. × 550.

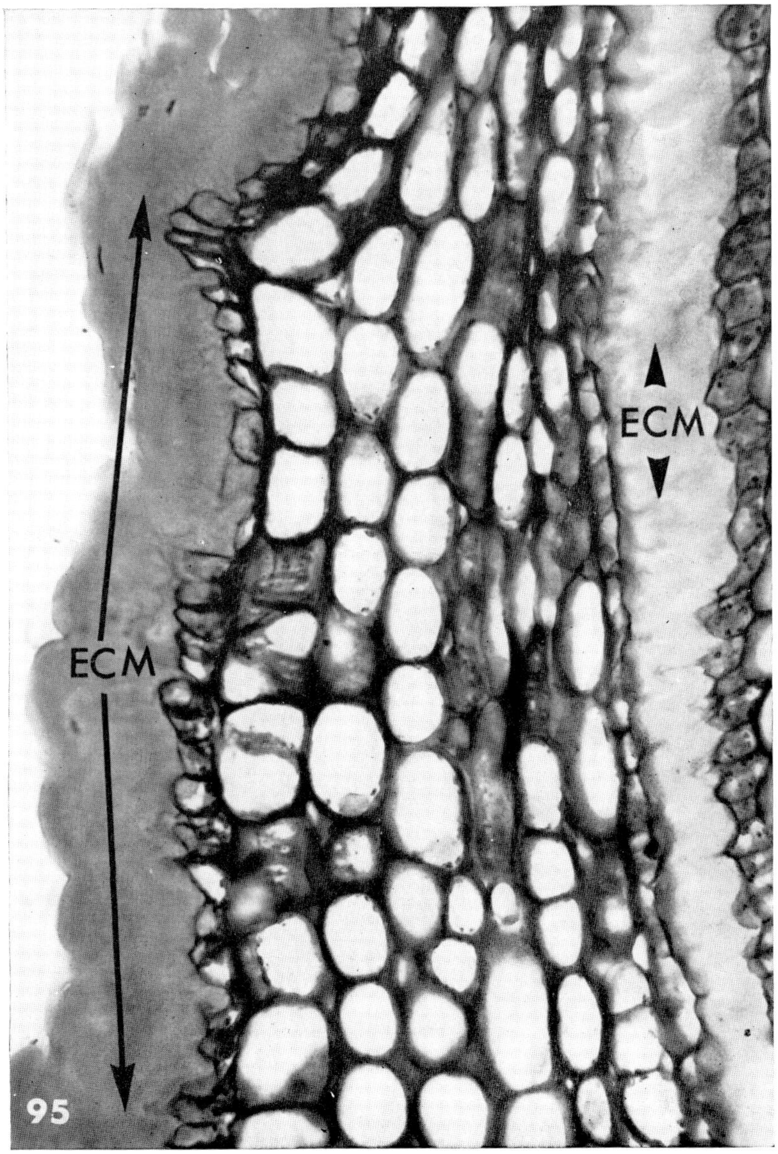

allows the closest pair of lateral buds to develop into two equal shoots (see Figure 57).

It has been shown that the application of a solution of a cytokinin directly to inhibited buds of many herbaceous plants will release them from dormancy even if the main shoot apex is left intact (*6.3, 6.4, 6.6, 6.7*). Much of this work has been done with peas. Figures 96 and 97 allow you to compare the appearance of a dormant lateral bud at the second node of a young pea plant with a similar bud that has just been released from dormancy by direct application (three days previously) of a cytokinin solution. While the bud is in the dormant state it has some xylem and phloem elements differentiated within its own axis, but it is linked to the vascular system of the stem only by provascular tissue (Figure 96). Three days after the application of the cytokinin, however, the bud is joined to the vascular system of the stem by xylem and phloem which have differentiated from the provascular tissue. A similar link of vascular tissue is formed if the dormancy of the bud is broken by the decapitation of the main axis. Whether the completion of the vascular link between stem and bud is a cause or an effect of the breaking of dormancy remains to be clarified (*6.8*).

There is a noticeable change in the apical meristem of a lateral bud of pea when its dormancy

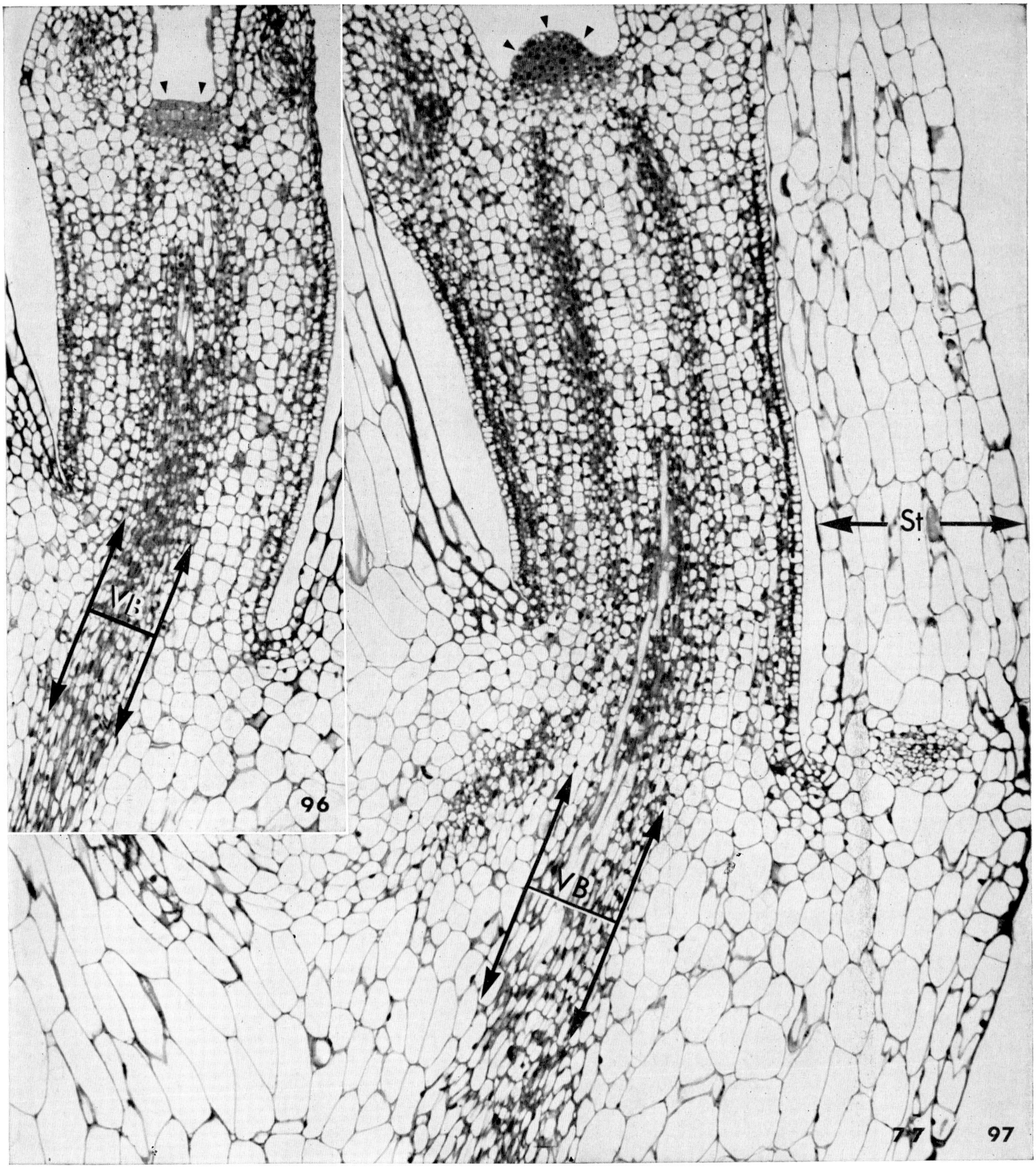

96 Longitudinal section through a dormant lateral bud at the second node of a pea stem showing its provascular connection (VB) with the stem and its apex (small arrows). PAS/toluidine blue. × 160.

97 Section similar to that of Figure 96 but of a laterial bud released from dormancy by the direct application of kinetin 3 days previously. The enlarged shoot apex (small arrows) is now apparent. St, stipule. PAS/toluidine blue. × 160. (Tissues kindly prepared by Dr. Tsvi Sachs.)

98 LS of dormant lilac bud showing raphides. Polarized light. × 550.

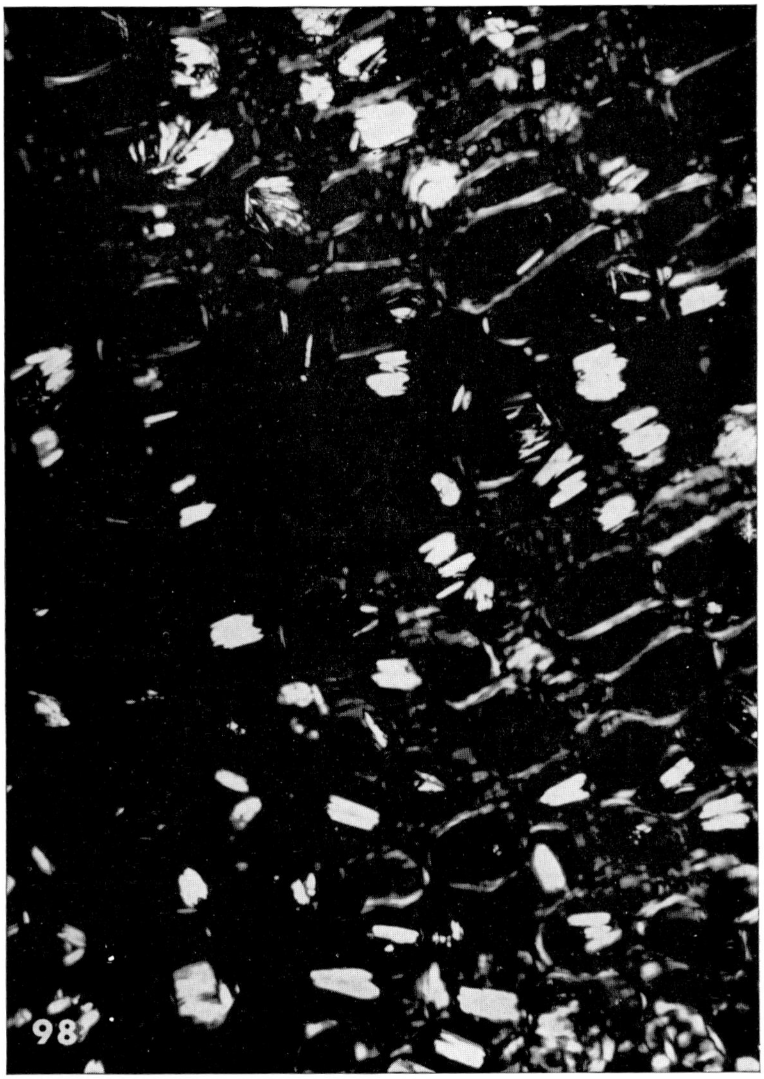

is broken. The meristem increases in volume and changes from a flat shape (Figure 96, arrows) to a conical shape (Figure 97, arrows). The breaking of dormancy is also marked by an increase in cell size throughout the young shoot and by the appearance of mitotic figures especially in the apical regions. At the cellular level, however, few differences have been detected between the components of the cells of dormant buds and those of corresponding cells of buds after dormancy has been broken. An exception to this is the marked increase in the frequency of occurrence of *raphides* (in some cases known to be composed of calcium oxalate) in the vacuoles of cells of dormant buds (Figure 98) compared to those of growing apices. Since raphides occur in a variety of cell types, it is difficult to assess their significance in the maintenance of bud dormancy.

GENERAL REFERENCES

Clowes, F. A. L. *Apical Meristems.* Oxford: Blackwell Scientific Publications, 1961, Chap. 9.

Esau, K. *Plant Anatomy.* 2d ed. New York: John Wiley & Sons, 1965, Chap. 5.

ROMBERGER, J. A. "Meristems, Growth and Development in Woody Plants," U.S. Department of Agriculture Technical Bulletin No. 1293, 1963.

WARDLAW, C. W. *Organization and Evolution in Plants.* New York: Longmans, Green & Co., 1965, Chap. 10.

REVIEWS AND RESEARCH PAPERS

6.1 EAGLES, C. F., AND P. F. WAREING. "The role of growth substances in regulation of bud dormancy," *Physiol. Plant.*, 17 (1964), 697–709.

6.2 CORNFORTH, J. W., et al. "Identity of sycamore 'dormin' with abscisin II," *Nature*, 205 (1965), 1269–1270.

6.3 PANIGRAHI, B. M., AND L. J. AUDUS. "Apical dominance in *Vicia faba*," *Ann. Bot.*, 30 (1966), 457–473.

6.4 SACHS, T., AND K. V. THIMANN. "Release of lateral buds from apical dominance," *Nature*, 201 (1964), 939–940.

6.5 SNOW, R. "On the nature of correlative inhibition," *New Phytol.*, 36 (1937), 283–300.

6.6 WICKSON, M., AND K. V. THIMANN. "The antagonism of auxin and kinetin in apical dominance," *Physiol. Plant.*, 2 (1958), 62–74.

6.7 GARRISON, R., AND R. H. WETMORE. "Studies in shoot-tip abortion, *Syringa vulgaris*," *Amer. J. Bot.*, 48 (1961), 789–795.

6.8 SOROKIN, H. P., AND K. V. THIMANN. "The histological basis for inhibition of axillary buds in *Pisum sativum* and the effects of auxins and kinetin on xylem development," *Protoplasma*, 59 (1964), 326–350.

99–102 Free-hand sections of fresh material stained with toluidine blue. For a summary of the color reactions of this dye, see the Appendix.

99

100

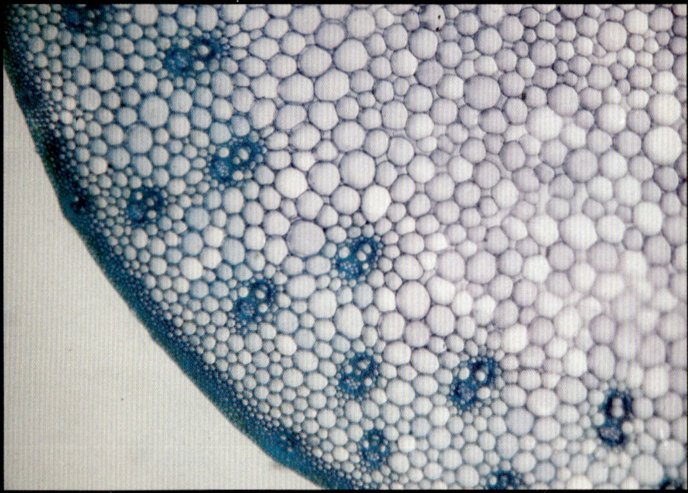
101

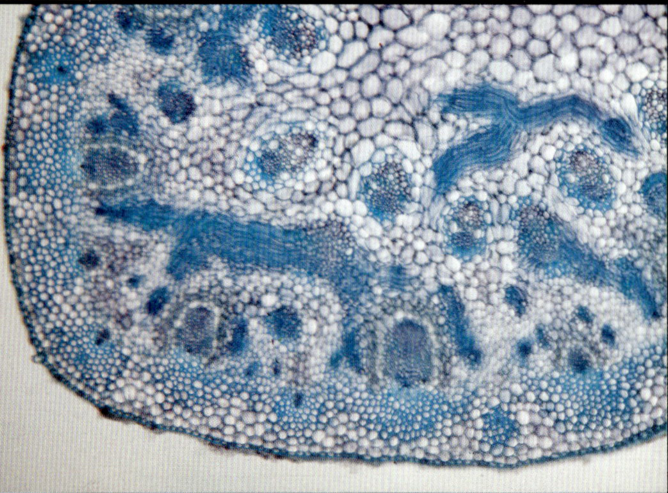
102

99 Transverse section of part of a pea stem showing a vascular bundle with a prominent cap of phloem fibers (stained bright blue). Cambial activity has just begun and interfascicular xylem and phloem are being produced. The chlorenchyma can be distinguished since chlorophyll is retained during preparation. × 105.

100 Transverse section of part of a young stem of *Hebe* spp. Although rapid secondary growth was occurring in this stem as indicated by the broad (unstained) cambial zone, the primary tissues (epidermis, cortex, and central parenchyma) are still intact. Note the fibers (bright blue) in the outer layers of the phloem. × 62.

101 Transverse section of an internode of orchard-grass (*Dactylis glomerata*), showing the vascular bundles and the peripheral ring of lignified sclerenchyma. × 52.

102 Transverse section at the node of a *Dactylis* stem showing the complex interconnections between the vascular bundles. × 52.

7

The Stem

IN CHAPTER 4 the shoot apex was examined as the site of leaf initiation, and it should be clear that in angiosperms the initiation, growth, and differentiation of the tissues of the leaves and stem are highly correlated. The stem develops into *nodes* (the regions at which the leaves are attached) and *internodes*. The elongation of the stem, which is influenced by a variety of factors, is largely due to extension of the internodes. In some plants, the shoot system consists of *long shoots* and *short shoots;* the internodes of the latter scarcely elongate.

The precise pattern of cell division, growth, and cell differentiation of developing nodes and internodes varies widely among species and may vary throughout the life of any one stem. Internodes may elongate acropetally (towards the apex) or basipetally (towards the root). The internodes may elongate in sequence, or the growth periods of several internodes may overlap. The extent to which cell division accompanies growth of the stem both in length and width is also variable. Just as the multiplicity of tissue patterns of mature leaves tends to obscure the importance of the photosynthetic parenchyma, so these variations in the development of stems tend to obscure the significance of the vascular system. In plants with well-developed leaves, the stem functions primarily as an organ of conduction, providing the pathway that connects the photosynthetic parenchyma of the leaves with the absorptive tissues of the root. It is not surprising, therefore, that the provascular tissue of stems develops in relation to the growing leaf primordia, and that differentiation of mature xylem and phloem elements within this provascular tissue occurs both in the stem and in the young leaves.

Although conduction may be its most important function, the stem does serve other functions. Photosynthetic parenchyma is commonly present, and the pith (when present) and extrafascicular parenchyma are clearly important areas for storage. In addition, the stem is an organ of support, and collenchyma and sclerenchyma are well-developed. As with the leaf, the epidermis protects the underlying tissues and regulates water loss and the exchange of other gases; commonly, the epidermis has stomata and glands or hairs, though these are often less abundant on the stem than on the leaf of the same species.

A detailed treatment of the different types of vascular systems, the differentiation of their provascular tissues, and the subsequent sequences of cellular differentiation which form the xylem and phloem are not illustrated here, for they are beyond the scope of this book. Excellent treatments of this subject are available elsewhere (see general references at end of this chapter). We wish to illustrate only the major features of two rather different systems, discuss briefly the initiation of the cambium, and to focus chiefly upon the differentiation and function of selected elements of the xylem and phloem.

Figure 99 shows part of a transverse section of a pea stem, in which the vascular bundles are discrete and *collateral,* with xylem towards the center of the stem and phloem towards the outside. Each bundle is "capped" with a band of fibers, whereas the center of the stem consists of a large air space, derived from the *pith* as the stem increased in diameter.

The vascular bundles of grasses are also collateral but smaller and more numerous than those of the pea stem. Figure 101, a transverse section of the stem at an internode of orchard-grass, illustrates an arrangement of tissues characteristic of many grasses. Here, the vascular bundles are in two rows close to the circumference of the stem, whereas the sclerenchyma forms a ring just under the epidermis. Figure 102 illustrates a transverse section through part of a node of the same stem shown in Figure 101. Nodes are the sites of much anastomosing and cross-linking of vascular strands; some of the complexity of these interconnections is apparent in Figure 102.

Secondary Growth

The vascular cambium is initiated in the provascular tissue that lies between the xylem and the phloem (Figures 99, 100, 103). The cell divisions are remarkably regular, forming files of cells that differentiate into phloem elements towards the outside of the stem and xylem elements towards the inside of the stem. In plants with discrete vascular bundles (for example, the pea seedling) this wave of division, which starts in the vascular bundles, spreads laterally from the bundles into the interfascicular parenchyma and forms a complete cylinder of meristematic cells. This new meristem is termed a *lateral meristem* to distinguish it from the apical meristems of the shoot and root. The most

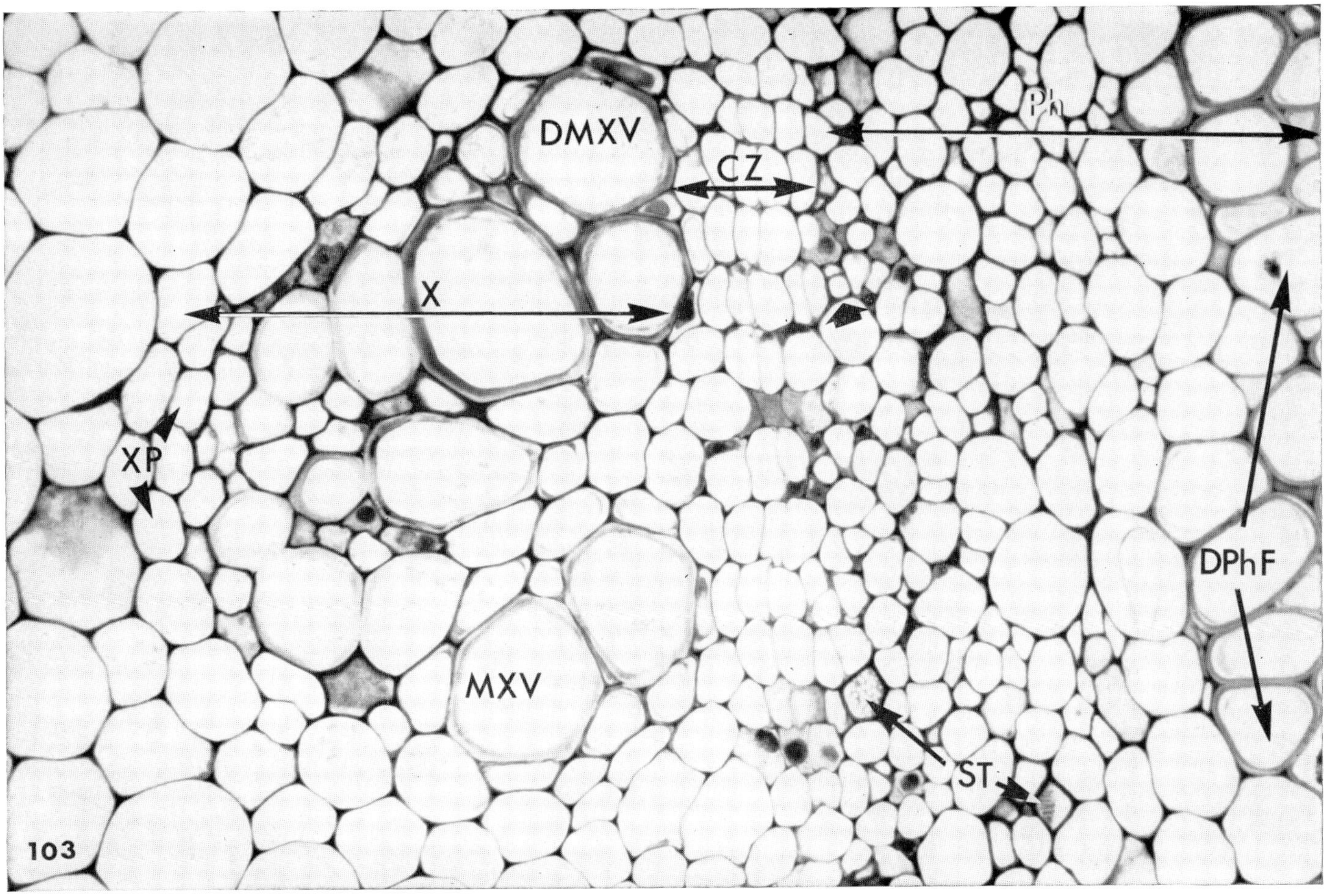

103 A transverse section of a vascular bundle in the stem of pea, showing the arrangement of the mature primary tissues and the development of the vascular cambium. Note the sieve element and companion cell (arrow) which have just differentiated from the cambial derivatives. CZ, cambial zone; DMXV, differentiating metaxylem vessel; DPhF, differentiating phloem fiber; MXV, metaxylem vessel; Ph, phloem; ST, sieve tube; X, xylem; XP, xylem parenchyma. Toluidine blue. × 620.

frequent divisions of the cambial initials are periclinal, generating radial files of cells on both sides of the cambial zone.

In a herbaceous plant like the pea, the amount of tissue formed by cambial division is limited, but in woody plants secondary growth continues so that xylem and phloem produced by the cambium become the major components of the stem. Figure 100 is a transverse section of part of a young stem of *Hebe* spp. in which the products of secondary growth are beginning to dominate the stem. The primary tissues of the epidermis and cortex are still intact at this stage. Later these primary tissues will be crushed and obliterated as more secondary tissue is formed.

The cambial zone contains two main types of meristematic cell, the *fusiform initials* and the *ray initials.* The former are elongated parallel to the axis of the stem and their derivatives form tracheary elements or fibers in the xylem or sieve tube mother cells and fibers in the phloem. The ray initials are not elongated appreciably in the long axis of the stem and form parenchyma cells of the xylem and phloem rays. The factors that regulate the precision of the planes of division in the cambium and the pathways of cellular differentiation upon which cambial derivatives pass are no better understood than the control mechanisms that operate in the apical meristems. The elucidation of these problems constitutes one of the central challenges in plant morphogenesis.

Differentiation of Xylem: Tracheary Elements

At maturity, tracheary elements consist only of their cell walls; all of the nuclear and cytoplasmic structures are hydrolysed and the lumen of the

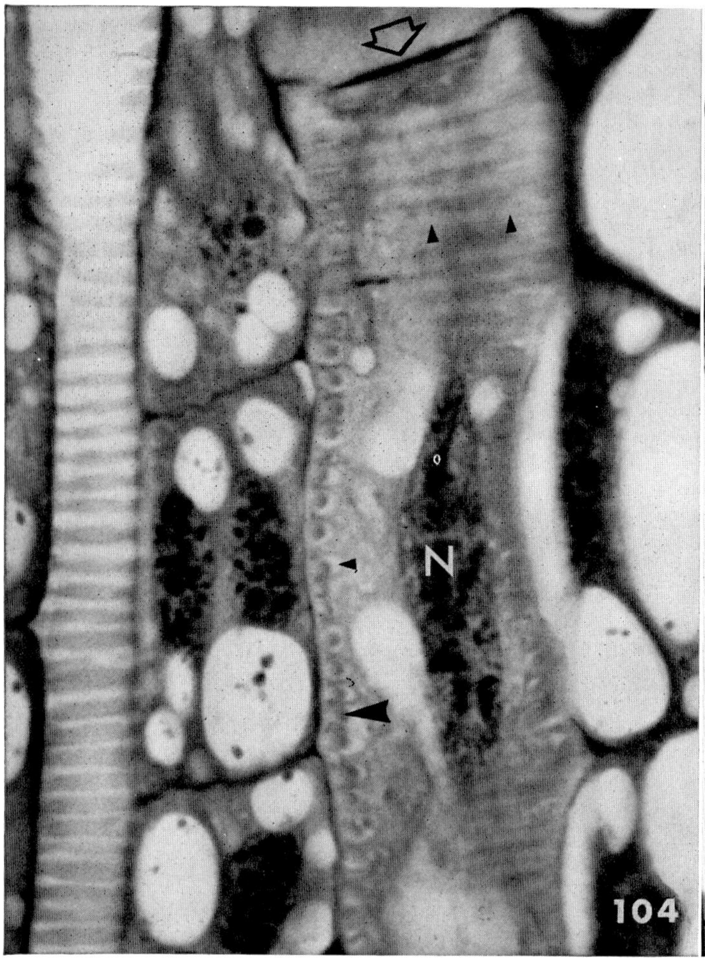

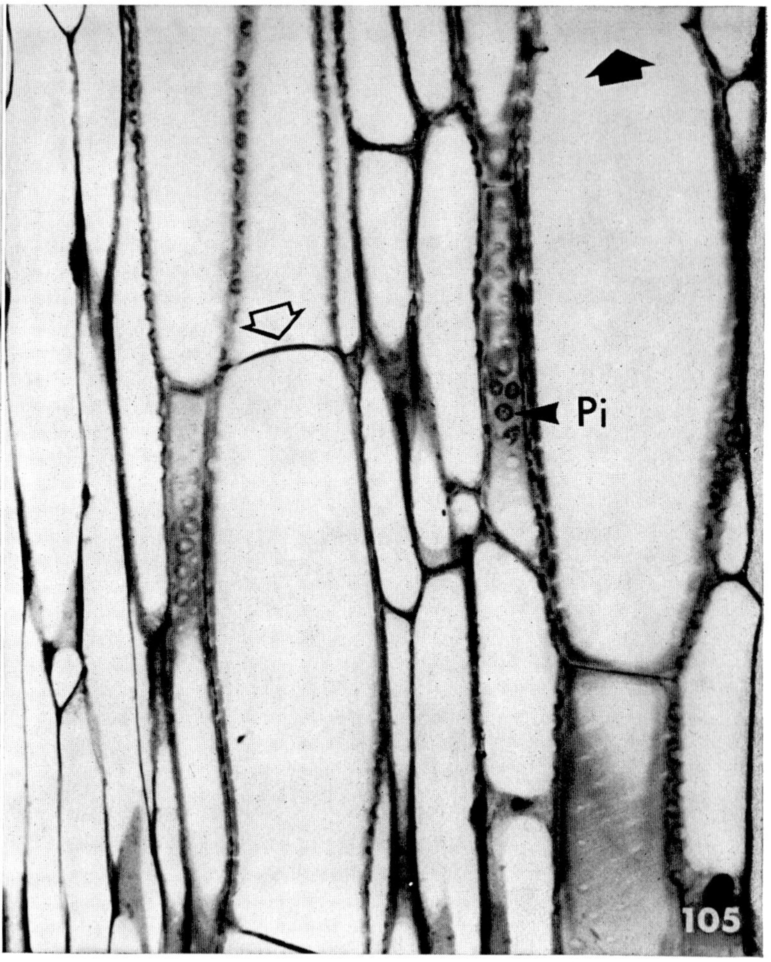

104 A tracheary element in the stem of pea. The bands of thickened wall, which give the mature element its characteristic appearance (see mature tracheary element on the left), are being deposited in the tracheary element on the right. At this stage, the bands of thickening consist of an unlignified outer region (little arrows) and a lignified inner region (large solid arrow; see also Figure 106). The open arrow shows a cross wall with its characteristically thickened (but unlignified) middle section. N, nucleus. Toluidine blue. × 1940.

105 Mature tracheary elements in the stem of pea. In some elements, the cross walls have broken down (solid arrow) to form vessels; in others (open arrow), the cross wall is still intact. These tracheary elements are from a node and the lateral walls are thickened everywhere except at the pit fields (Pi), seen in face view. Contrast this pattern with the annular banding shown in Figure 104. Toluidine blue. × 1940.

106-107 Stages in the differentiation of tracheary elements in a leaf of *Phaseolus vulgaris.*

106 This cell is still depositing the wall thickenings that have become partly lignified. Note the abundance of dictyosomes (D), rough endoplasmic reticulum, (ER) and mitochondria (M) during this stage of wall synthesis. Bundles of microtubules (visible at higher magnification; not illustrated) lie above the bands of developing wall. LW, lignified wall; P, plastid; UW, unlignified wall; V, vacuole. × 13,400.

107 A later stage showing collapse of the cytoplasmic structures (part of this cell is seen at lower magnification in the lower right hand corner of Figure 106). The cell membrane (CM) has begun to withdraw from the wall, which has not yet been hydrolyzed in the unlignified layers between the bands (compare with Figure 109). LW, lignified wall. × 32,000.

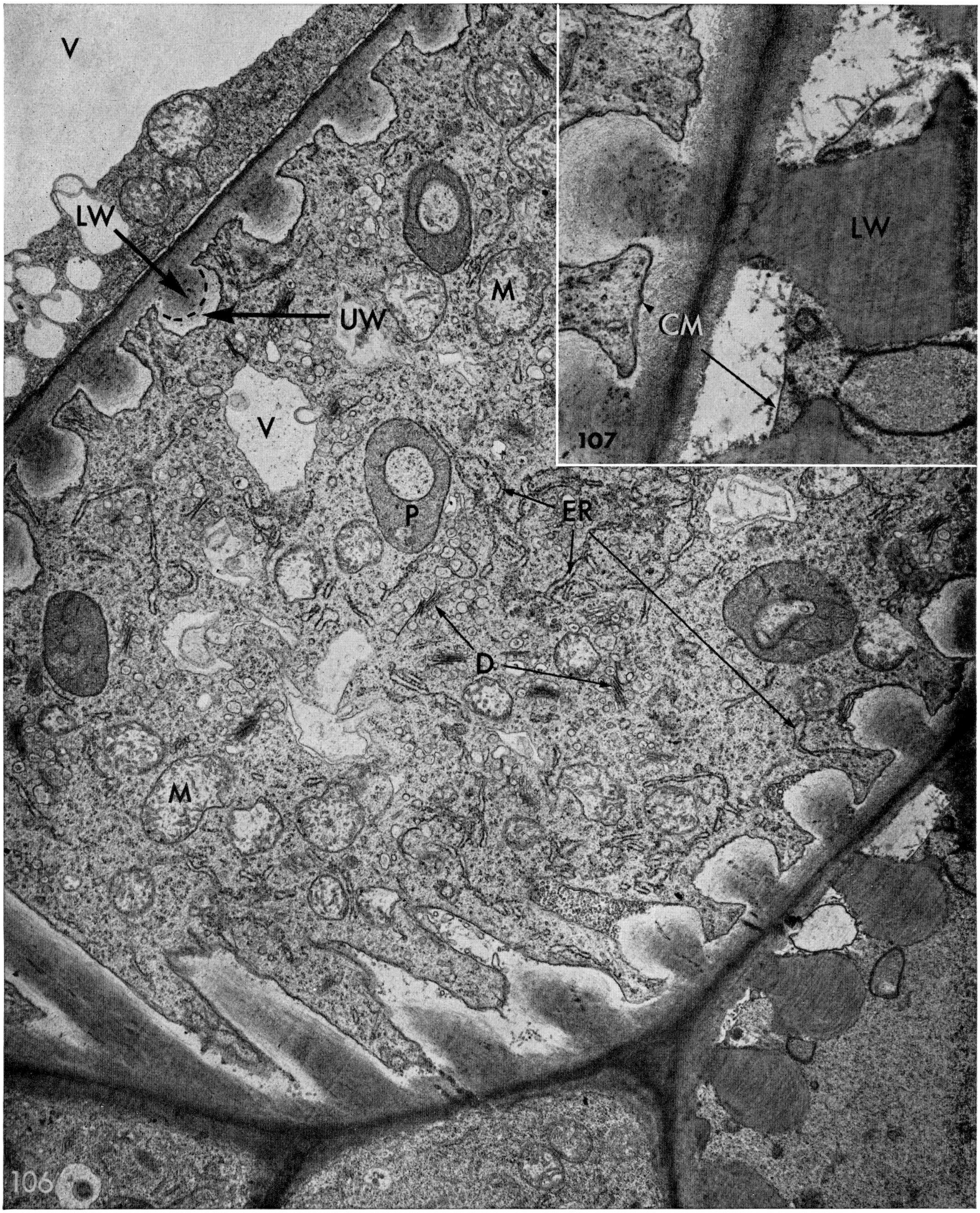
V
LW
UW
M
V
P
ER
D
M
LW
CM
107
106

former cell becomes filled with an aqueous solution, the transpiration stream. In thinking about the structure of tracheary elements, it is helpful to bear this mature state in mind. These cells are the instruments by which the plant body constructs a complex, internal water reticulation system, derived solely from carefully "engineered" extracellular material. Tracheary elements are classified into two types, *tracheids* and *vessel members.* These differ in the mature state in that, in tracheids the lumina of adjacent elements are connected only through relatively small perforations at pits, whereas the lumina of adjacent vessel members are connected in at least one area (commonly at their ends) by a large perforation (Figure 105). In some cases, this perforation may extend over the complete diameter of a vessel member. A vertical file of such interconnected vessel members is called a *vessel.*

The pattern of thickening of the wall of tracheary elements is governed by the conditions of growth that prevail during differentiation of the elements. In the protoxylem in general, and especially where rapid extension accompanies differentiation, the wall thickenings take the form either of annular bands (Figure 104) or of helices. Where differentiation is accompanied by a minimum of extension (as in secondary xylem or in primary xylem of nodes), the thickening occurs over the whole surface of the lateral walls except at the pits (Figure 105). The thickened regions of these walls are invariably lignified, but in the primary xylem, the *compound middle lamella* is not lignified, even beneath the lignified bands of thickening (Figure 108). In the secondary xylem, lignification usually extends to this region. In vessel members, the site of the future perforation is not lignified even though the primary wall may thicken (Figure 105, open arrow). Perforation is effected by the dissolution of this unlignified thickened wall.

The development of these lignified wall thickenings has been studied in some detail during the past few years (*7.1, 7.2*). Figure 104 is a light micrograph of a differentiating vessel member in the protoxylem and Figures 106 and 107 show two stages in the differentiation of similar elements as seen with the electron microscope. Figure 106 shows the characteristically rich cytoplasm of these differentiating tracheary elements, with its wealth of endoplasmic reticulum and dictyosomes. Note the elongated nucleus, with large chromatin aggregates, and the areas of intense basophilia near the unlignified cross wall (Figure 104). In surface view, the bands of thickening are lightly stained (Figure 104, little arrows); between these bands lie more densely stained areas, which correspond to regions rich in rough endoplasmic reticulum in Figure 106. When the bands are seen in section (Figures 104 and 106), it is clear that the lignin is deposited first in the earliest-formed part of the thickening. It has been demonstrated in several different tissues that the distribution of microtubules (*7.2, 7.3*) is correlated with the development of these bands of thickening. When thickening begins, bands of microtubules run circumferentially around these cells at right angles to the long axis of the cell. The bands of thickening form first *between* these bands of microtubules, but as thickening progresses, microtubules come to lie above the thickenings. It is speculated that the microtubules channel dictyosome vesicles to the sites where the wall is thickening just as they seem to do in the developing cell plate (Figures 35 and 36). Further support for this view comes from experiments with colchicine, which have shown that in cells treated with this drug, wall material is still deposited but the precise pattern is lost. Microtubules cannot be identified in the colchicine-treated cells (*7.4*).

The total loss of the cytoplasmic and nuclear contents is a most dramatic event, one that has not been studied in detail. The cell in the lower right-hand corner of Figure 106 was fixed during this period of cytoplasmic destruction. Only a few internal membranes, probably derived from the endoplasmic reticulum, are still discernible in the cell, and the cell membrane has just begun to separate from the wall (Figure 107). It has been suggested, but not yet proved, that hydrolytic enzymes are released into the cell and bring about this destruction of its contents.

If one examines the unlignified walls of mature tracheary elements from the primary xylem, it is clear that they too are modified during the final stages of differentiation. As Figure 108 shows, the primary wall fails to stain with cationic dyes in the regions between the lignified bands of thickening. In the electron microscope, this region is seen to consist of just a few fibrils, which may be cellulose and/or protein. The noncellulosic polysaccharides, including the polyuronides which stain with the cationic dyes, have been dissolved. They are not dissolved in the regions that are lig-

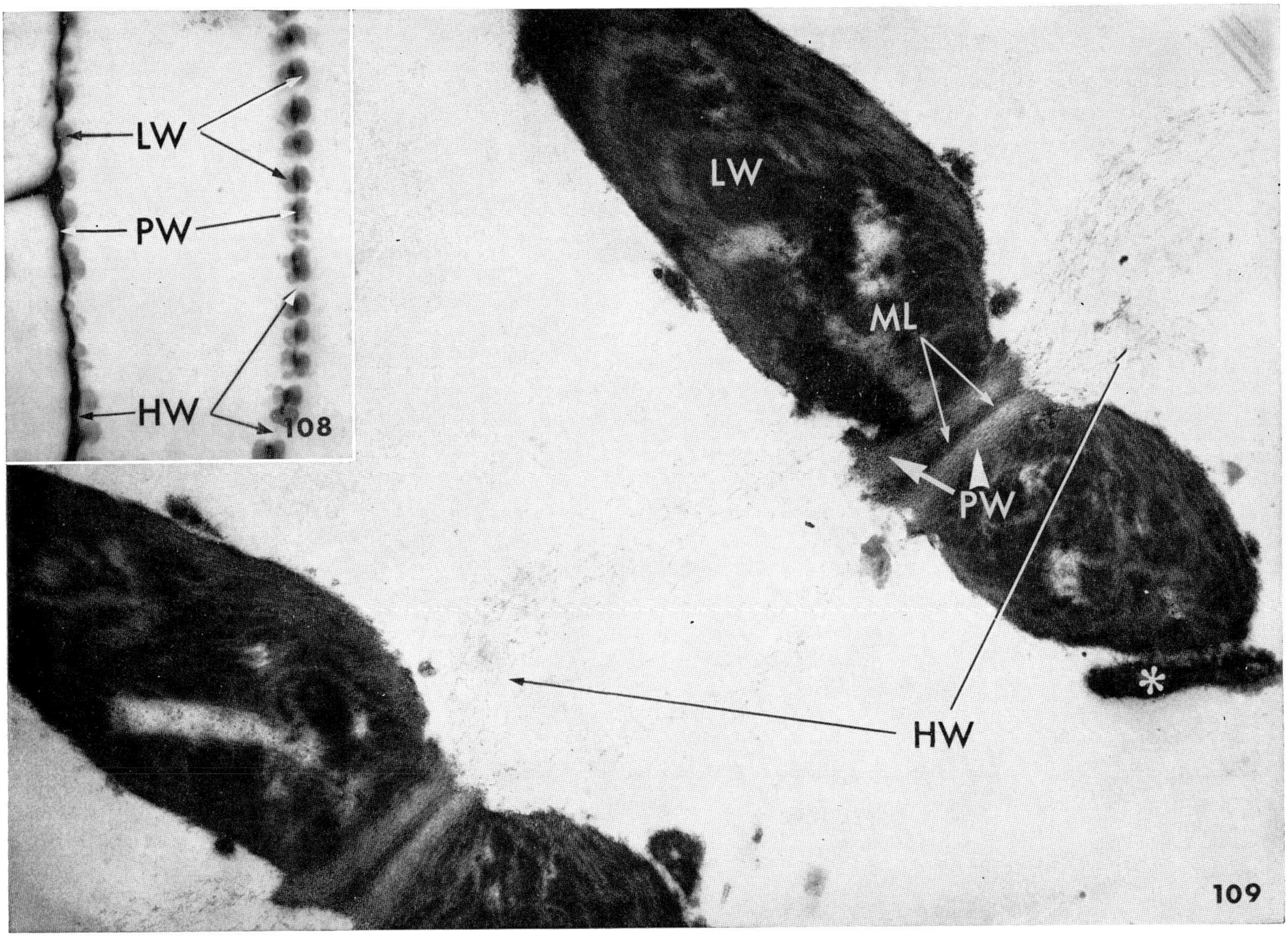

108-109 Hydrolyzed walls in tracheary elements.

108 Longitudinal section through protoxylem tracheary elements in a pea stem. Note the intense staining of the unlignified primary wall (PW) in the regions that underlie the lignified bands of thickening (LW). This staining reaction does not occur in the regions *between* the bands marked HW. Toluidine blue. × 1420.

109 Electron micrograph of a similar region in the protoxylem of an oat coleoptile. The primary wall (PW) and middle lamella (ML) are seen to be intact in the regions that underlie the bands of lignified thickening (LW), but consist only of fine fibrils in the hydrolyzed region of the wall (HW). Cytoplasmic remnants at asterisk. × 36,400. (Reproduced with permission from *Protoplasma,* 63, 1967, p. 451.)

nified nor in the regions that underlie the lignified thickenings of the wall (Figures 108 and 109). Nor does hydrolysis of wall polysaccharide proceed beyond the middle lamella where a tracheary element abuts a parenchyma cell or any other living cell in the vascular bundle (*7.5*).

These various modifications of the extracellular material during the final stages of differentiation of tracheary elements help to explain several features of the xylem. Firstly, wall hydrolysis is clearly the mechanism by which vessels are formed from vessel members. Secondly, one may understand the full significance of annular rings of thickening in the rapidly growing protoxylem. These cells differentiate during rapid extension and will be stretched passively after differentiation by the continuous extension of the cells around them. An unlignified wall cannot be extended passively for more than a small percentage of its length without rupture. How the cytoplasm prevents rupture during elongation of several hundred percent is still one of the central mysteries of plant cell growth (*7.6*), but it clearly cannot do so once it has been killed. It would appear that

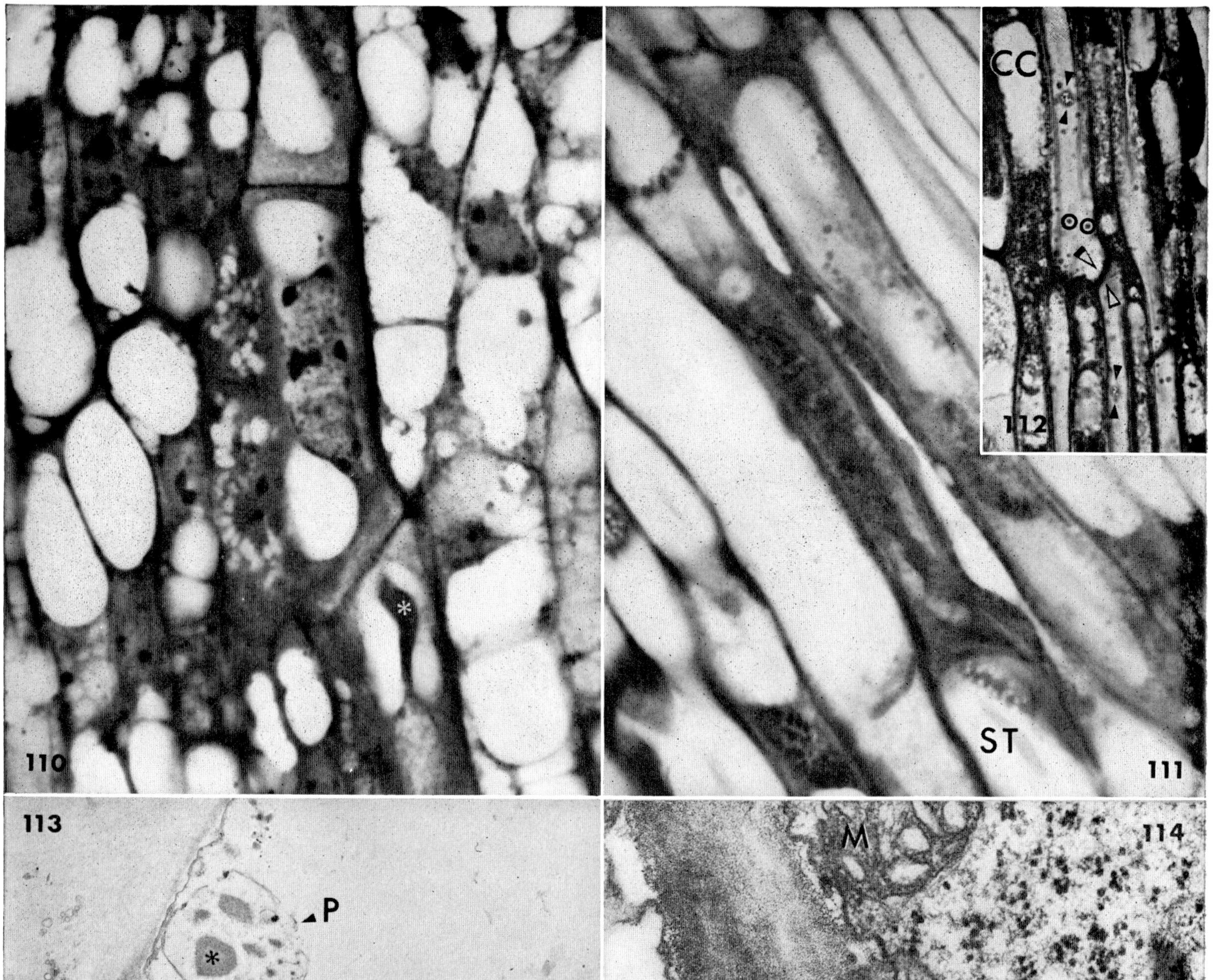
110
111
CC
112
ST
113
P
P
114
M
ER
ST

hydrolysis of the noncellulosic polysaccharides from the unlignified parts of the walls removes the resistance of the wall as a whole to stretch at just the time at which the cytoplasm is no longer capable of doing the same task. The lignified bands survive, and because they are anchored to the walls of the surrounding cells and in some cases to one another, they cannot tilt, ensuring that the lumen of the "pipe" remains open even though the pipe is being stretched considerably.

Differentiation of the Phloem

There is no doubt that the phloem contains the cells that carry out the bulk of long-distance transport of sugar in the intact plant. Unfortunately there is no completely convincing, direct evidence that this transport takes place within the *sieve tubes* in the intact plant, although these cells are certainly the most likely sites of this activity. At the present time, there is no general agreement either about the mechanism of the transport process (even assuming that it occurs in sieve tubes) or about the structure of the sieve tubes at the time when they are engaged in transport (*7.7, 7.8, 7.9*). Since, in the minds of many workers, arguments about the mechanism of transport are intimately related to arguments about structure of sieve tubes, there are serious controversies in the field (*7.10, 7.11, 7.12*). An examination of some of the stages in differentiation of sieve tubes will give an appreciation of the difficulties involved.

As with the formation of tracheary elements, the differentiation of sieve elements is accompanied by a modification of the wall and cytoplasm of the provascular cells. Several of the changes take place concurrently, but the details vary among species and even among different vascular bundles of the same plant (*7.13*). In most cases, the lateral walls become somewhat thicker than those of nearby parenchyma cells, whereas the cross walls begin to form *sieve plates.* The cytoplasm may develop protein-rich bodies of various shapes called "slime bodies," and the plastids often develop paracrystalline inclusions. Various stages in this process are shown in Figures 110 to 115.

110-114 **Stages in the differentiation of sieve elements.**

110 **Slime body formation (asterisk) and a very early stage in the modification of the cross walls which form sieve plates. The sieve element contains a prominent nucleus at this stage. *Tropaeolum* stem. Acid fuchsin/toluidine blue. × 1940.**

111 **Late stage in sieve element (ST) differentiation in *Tropaeolum*. The cross walls are now strongly modified but the nucleus and cytoplasm are still intact. Acid fuchsin/toluidine blue. × 1940.**

112 **Mature sieve elements in a wheat coleoptile. Note the sieve plate (large arrows), sieve areas on the lateral walls (small black arrows), and the characteristic sieve element granules of grasses (circled). Nuclei are absent. CC, companion cell. Acid fuchsin/toluidine blue. × 735.**

113 **Electron micrograph of adjacent sieve elements in different stages of differentiation in an oat coleoptile. In the lower elements, the cytoplasm is intact but the plastids have begun to differentiate the crystalloids (asterisks) characteristic of the sieve element plastids (P). These modified plastids are the sieve element granules of Figure 112 (see also Figure 115). Note the electron transparent areas of wall around the plasmodesmata (open arrows) of the future sieve plate. This stage would correspond approximately to that shown in Figure 110. × 16,600.**

114 **Electron micrograph of a sieve area between a phloem parenchyma cell and a sieve element (ST). On the parenchyma cell side, the structure of the plasmodesmata (white arrow) is essentially the same as that seen in other parenchyma cell walls (see Figures 7 and 8). On the sieve element side, the wall is modified around the plasmodesmata (asterisk) in a manner similar to that shown at the sieve plates (compare with Figures 113 and 115). ER, endoplasmic reticulum; M, mitochondria. × 57,600. (Figures 113 and 114 reproduced with permission from *Protoplasma*, 63, 1967, pp. 461 and 463.)**

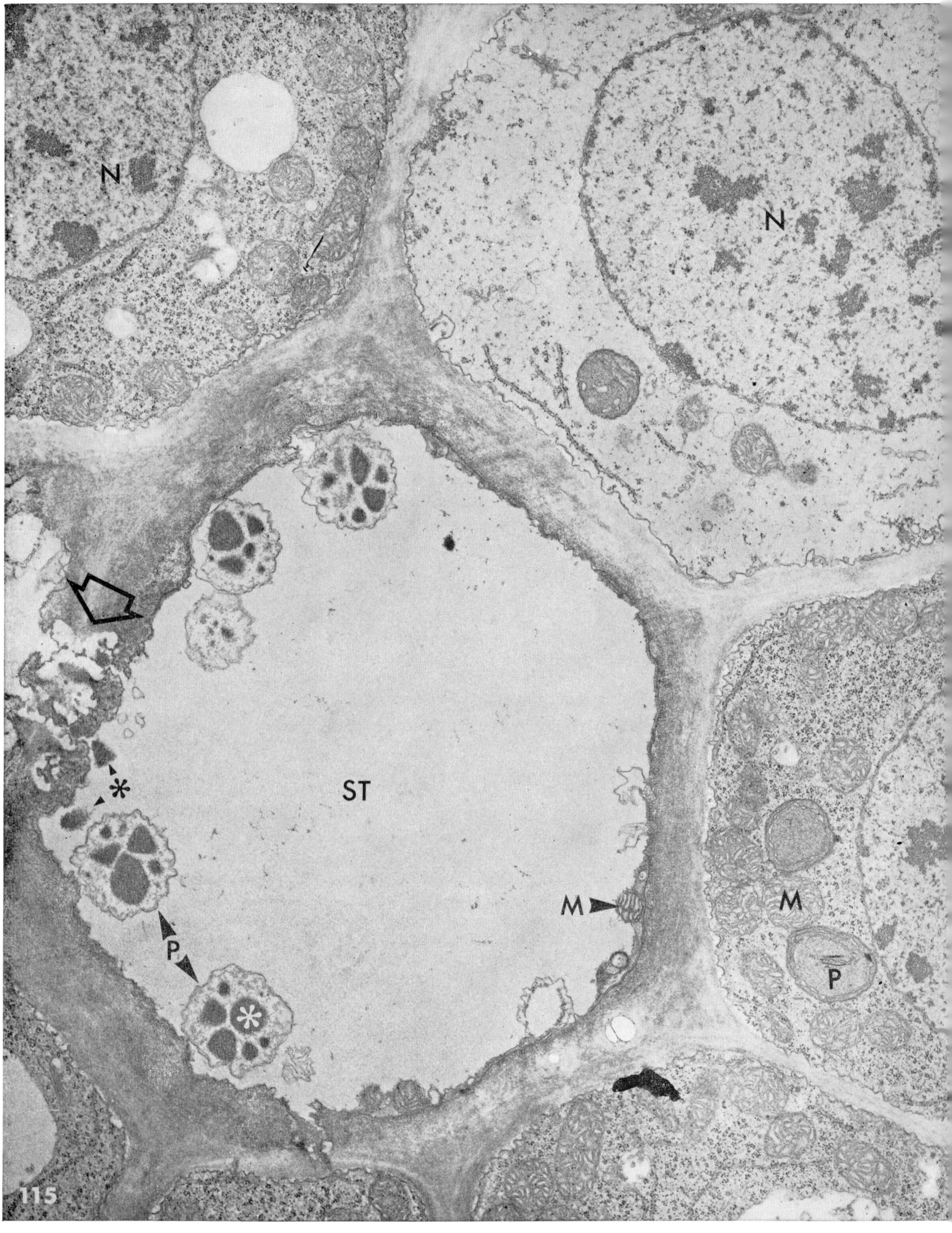
N
N
ST
*
*
M
M
P
P
115

The differentiating sieve elements become extremely delicate quite suddenly during the course of these changes so that they become very difficult to fix. At about this time the nucleus disappears along with most of the "normal" cell organelles. Dictyosomes, which are often abundant during the stage of wall thickening, rough endoplasmic reticulum, and free ribosomes cannot usually be identified and the vacuolar membrane is no longer distinct. It is believed that cells in this state (Figure 115) are the ones primarily responsible for long-distance transport.

The formation of the sieve plate raises several interesting problems. In the light microscope, the future sieve plate begins to show regions that fail to stain with cationic dyes (Figure 110). As differentiation proceeds, the unstained areas enlarge (Figure 111). In the electron microscope, these unstained areas seem to correlate with zones of low electron density which form around the original plasmodesmata (Figures 113 and 115). Some workers believe that these unstained areas are filled with callose (*7.14*), but whatever the chemical composition of these areas may be, there can be no doubt that the original structure of the wall is modified during the development of these unstained areas.

In dead sieve tubes, which have lost all trace of cytoplasm, these unstained areas are replaced at least in part by true pores. In the mature elements, however, it is simply not known whether pores exist *in vivo,* or how wide they are if they do exist. In the living tissue, the sieve tubes develop a high hydrostatic pressure (as much as twenty atmospheres). When the permeability of the cell membrane is altered during fixation, this hydrostatic pressure is suddenly released, causing violent disorganization of the structure of the mature sieve tube contents (*7.15*). One cannot tell, therefore, whether the material that clogs the pores of the sieve plate in such preparations (Figures 114 and 115) was present *in vivo,* or whether it is a consequence of this disorganization. The very fact that the contents of the sieve tubes do tend to "pile up" at the sieve plates suggests that they may have open pores *in vivo.* The contents of ruptured plastids are often seen in the region of the plate in fixed material (Figure 115). It is possible that this material is responsible for the synthesis of wound callose, which, by blocking the pores, prevents further loss of sieve tube contents.

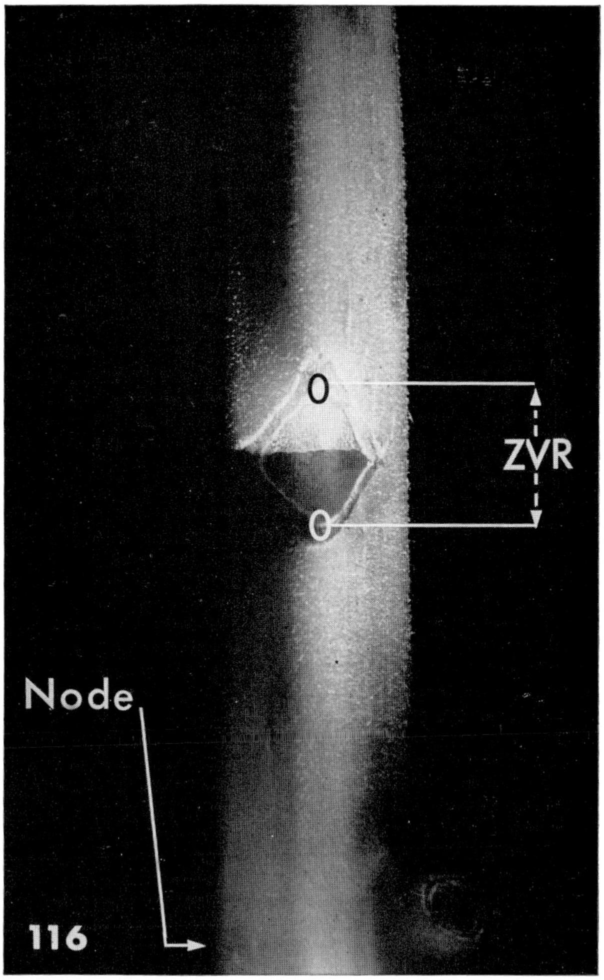

116 **A nick made in one corner of a *Coleus* internode. The ends of the severed vascular bundle are within the circles. The zone of vascular regeneration (ZVR) links the cut ends in the parenchyma tissue beneath the cut surfaces. × 5.5.**

115 **Electron micrograph of a transverse section of a mature sieve tube (ST) and adjacent parenchyma cells in the phloem of an oat coleoptile. The large arrow shows a sieve plate with the electron transparent material believed by many authors to be callose. Note that the sieve tube granules of Figure 112 are modified plastids with crystalloids (asterisk) which are readily released from ruptured plastids (P) when sieve tubes are cut. A few mitochondria (M) are evident in the mature sieve tube. N, nucleus; P, plastid. × 14,000. (Reproduced with permission from *Protoplasma,* 63, 1967, p. 457.)**

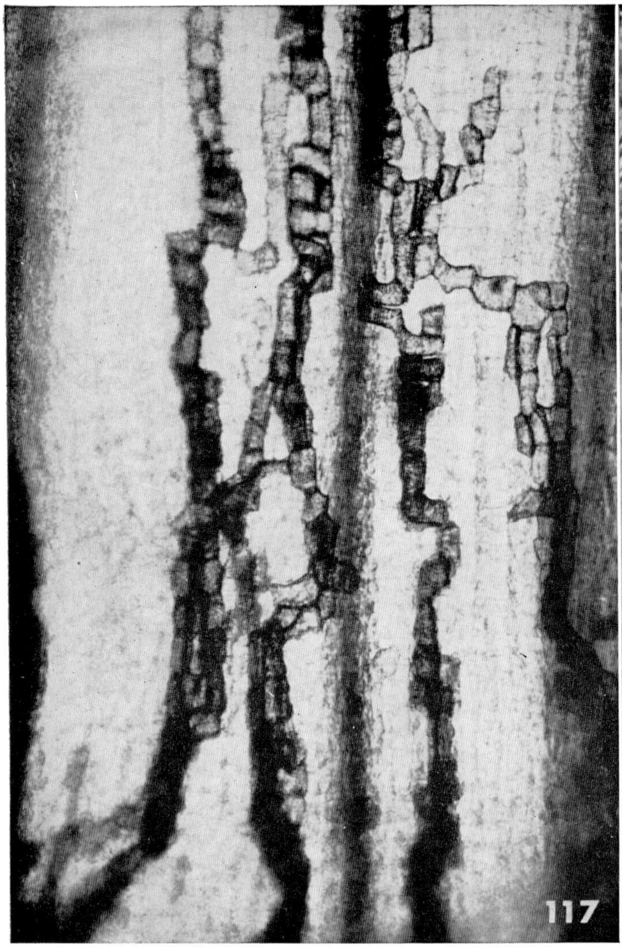

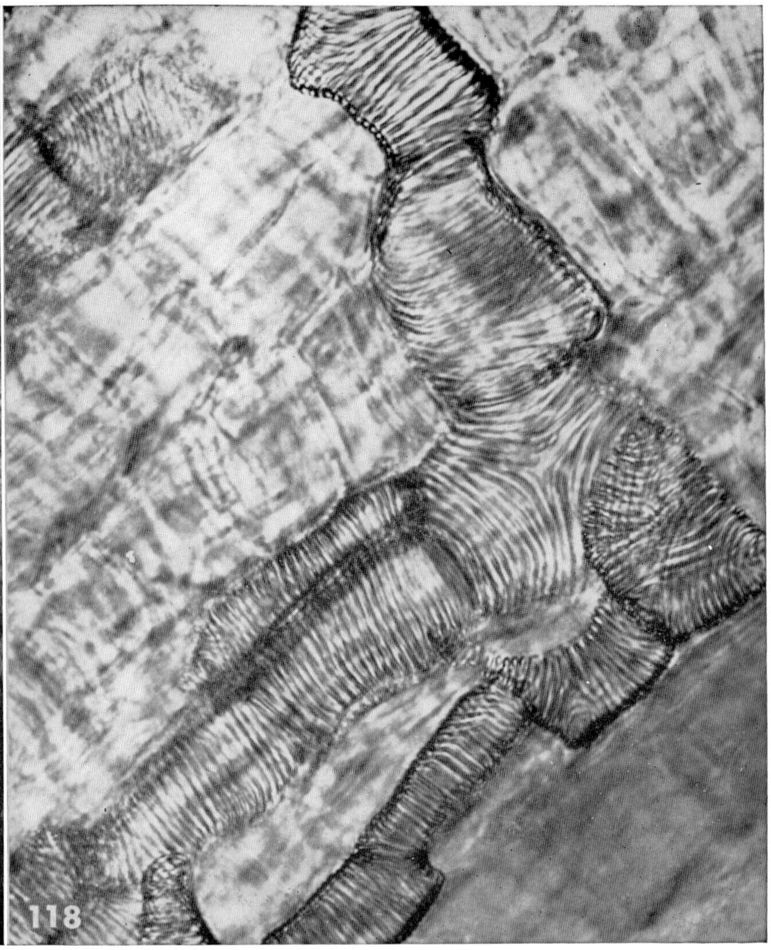

117 Tangential view of the xylem which has formed in the zone of vascular regeneration three days after a cut such as that shown in Figure 116 was made in the second internode of a young *Coleus* stem. Material cleared with lactic acid. Unstained. × 75.

118 Higher magnification of a portion of Figure 117. × 450. (Figures 117 and 118 courtesy Mr. J. Bell and Miss J. Millar.)

119 Longitudinal section of a nicked *Coleus* internode. Section passes through the center of the nick (which appears in the upper left corner) in a plane parallel to the regenerating vascular strand. The lower pair of arrows indicates the original vascular strand (the cut end is out of the plane of the section). Upper arrows show pathway along which a strand of provascular tissue is being formed by divisions of the parenchyma cells. The asterisks indicate early formed xylem cells (like those shown in Figure 117). Material fixed five days after the nick was made. PAS/toluidine blue. × 275.

120 Longitudinal section along a regenerating vascular strand at a later stage showing xylem formed from provascular cells (upper right), undifferentiated provascular tissue, and a newly formed sieve tube (DST). Material fixed seven days after the nick was made. PAS/toluidine blue. × 600.

121 Higher magnification of the sieve tube shown in Figure 120. × 1950.

122 Early stage in the differentiation of xylem elements from provascular tissue of a regenerating vascular bundle. The thickenings of the secondary wall are being laid down on the inner surface of the relatively thin, lateral primary walls (small arrows). No secondary wall is deposited on the end walls. The cytoplasm and nucleus (N) are still intact. Note the intense basophilia of the cytoplasm in the regions between the bands of thickening (large arrows). Compare with Figures 104 and 106. Toluidine blue. × 1780. (Sections for Figures 119–122 courtesy Mr. J. Bell and Miss J. Millar.)

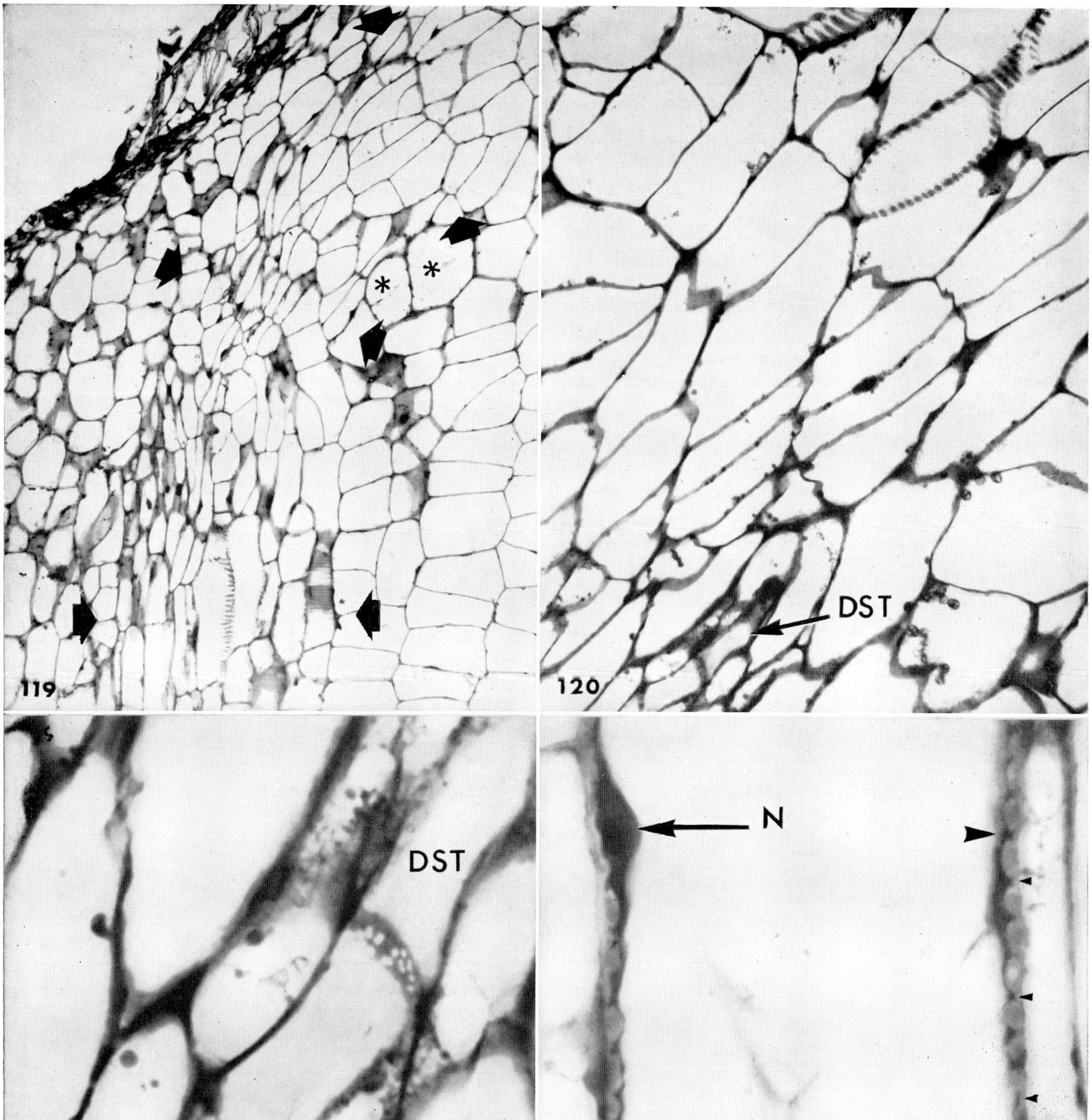
119
120
DST
121
DST
122
N

Modification of the lateral wall also occurs on the sieve element side where sieve elements have pit fields in common with adjacent parenchyma cells (Figures 112 and 114). These are termed *sieve areas*. These structures show that the influence that leads to wall modification almost certainly emanates from the differentiating sieve elements and that true sieve plates can form only between two differentiating sieve elements.

Although mitochondria do occur in mature sieve elements, they are much more abundant in the adjacent phloem cells, especially in companion cells (Figure 115). Companion cells often die at the same time as their associated sieve elements, and it is possible that these two cell types carry out complementary roles in the process of long-distance transport, the former providing the energy for the transfer of solutes to and from the latter.

The Regeneration of Vascular Tissues

During the normal development of a higher plant, the xylem and phloem elements differentiate either from provascular cells formed by root or shoot meristems, or from cambial derivatives (*7.16, 7.17*). Under certain abnormal conditions, however, vascular tissue can be induced to form from vacuolated parenchyma cells.

One of the best-known examples of induced vascularization occurs in the regeneration of severed vascular strands in stems (*7.18, 7.19, 7.20, 7.21, 7.22*). Figures 117 to 122 show some of the stages of vascular regeneration after one of the four major vascular bundles of the stem of a young *Coleus* plant was cut by a small wound (Figure 116) in the second internode below the shoot apex. Vascular regeneration in the material illustrated follows a definite sequence. A few days after wounding, the gap in the cut strand is bridged by xylem cells which form in a number of rather meandering files (Figure 117) that run behind the wound. These cells, unlike normal elongated xylem elements, are short and wide and appear to have differentiated directly from parenchyma without cell division (Figures 117 and 118) (*7.18*).

Later in the sequence there is localized cell division in the zone of regeneration, which produces around the wound (Figure 119) a strand of elongated cells that resemble provascular tissue. Some of these cells subsequently differentiate into xylem elements and others later form phloem (Figures 120, 121 and 122), so that the original vascular bundle is rejoined by both xylem and phloem.

Quite a different regeneration sequence may occur in other internodes of *Coleus*. For example, if the wound is made just below the fifth node (counting from the apex), the unusual xylem does not form and the first sign of regeneration is cell division that forms a provascular strand from which phloem and, later on, xylem are differentiated (*7.21*).

There is little or no regeneration of a cut vascular bundle of *Coleus* if the leaves above the wound are removed. However, if an appropriate concentration of auxin is applied to one of the petiole stumps, vascular tissues will regenerate completely, even if the whole top of the plant has been removed. It has been shown that most of the auxin moves basipetally in *Coleus* stems. Since the various stages of regeneration also appear to progress mainly in a basipetal direction, it has been suggested that they are mediated by the downward movement of hormone from the cut end of the vascular strand (*7.19, 7.22*) into the parenchyma of the zone of regeneration. Support for the idea that it is the cutting of the vascular tissue that initiates the regeneration comes from the observation that no vascular tissue regenerates in stems where only the parenchyma has been wounded.

Auxin has been shown to be necesary for the differentiation of vascular tissue from parenchyma in other situations—for example, in explants of tobacco pith (*7.23*) and in various tissue cultures (*7.24, 7.25, 7.26*). Although this hormone is undoubtedly necessary for regeneration in *Coleus*, diffusion of auxin along a concentration gradient from the cut end of the original vascular trace seems much too crude a device to mediate the highly oriented and regulated regeneration of the cut strand. Furthermore, one is faced with the observation that the steps in the regeneration can be in different sequence at different nodes, although the end result—a narrow bridge of vascular tissue between the ends of the severed bundle—is essentially the same in each case. How may one explain such diverse results? Some very interesting and important observations on a different system may help to provide the explanation. The differentiation of both xylem and phloem from parenchyma requires auxin, but the specific cell types formed depend upon the sugar concentration present (*7.25, 7.26*). With an appropriate concentration of auxin, high sugar levels favor the production of phloem, low sugar levels the production of xylem. Both cell types are produced with intermediate levels of sugar in the nutrient medium. It is quite possible that different sugar levels exist in *Coleus* internodes of different ages. Such inherent differences and their modification as regeneration progresses may explain the differences in the sequence of xylem and phloem regeneration.

At the cellular level the differentiation of the regenerating xylem and phloem cells seems to proceed in much the same way as the differentiation of these cells in the intact plant, even though the initial cells appear quite different. This is especially true in the case of the first-formed xylem in the younger nodes (Figures 117 and 118). In this case the initials are highly vacuolated parenchyma cells, which in the normal plant would have remained in this state. The regeneration stimulus, whatever its nature, releases them from their role as parenchyma and starts them down the dead-end path of xylem differentiation, apparently without even an intervening cell division.

Experimental systems like that of *Coleus* described here have helped to emphasize how much we have yet to learn about the mechanisms regulating cell differentiation in intact plants. Such systems have already provided some partial answers, but they offer much scope for further work.

GENERAL REFERENCES

BAILEY, I. W. *Contributions to Plant Anatomy.* New York: Ronald Press, 1954.

ESAU, K. *Plant Anatomy*, Chaps. 6, 10, 11, 12, and 15. New York: John Wiley & Sons, 1965.

———. *Vascular Differentiation in Plants.* New York: Holt, Rinehart & Winston, 1965.

ZIMMERMAN, M. H. *The Formation of Wood in Forest Trees.* New York: Academic Press, 1964.

REVIEWS AND RESEARCH PAPERS

7.1 CRONSHAW, J., AND G. B. BOUCK. "The fine structure of differentiating xylem elements," *J. Cell Biol.*, 24 (1965), 415–431.

7.2 PICKETT-HEAPS, J. D., AND D. H. NORTHCOTE. "The relationship of cellular organelles to the formation and development of the plant cell wall," *J. Exp. Bot.*, 17 (1966), 20–26.

7.3 HEPLER, P. K., AND E. H. NEWCOMB. "Microtubules and fibrils in the cytoplasm of *Coleus* cells undergoing secondary wall deposition," *J. Cell Biol.*, 20 (1964), 529–533.

7.4 PICKETT-HEAPS, J. D. "The effects of colchicine on the ultrastructure of dividing plant cells, xylem wall differentiation and distribution of cytoplasmic microtubules," *Developmental Biology*, 15 (1967), 206–236.

7.5 O'BRIEN, T. P., AND K. V. THIMANN. "Observations on the fine structure of the oat coleoptile. III. Correlated light and electron microscopy of the vascular tissues," *Protoplasma*, 63 (1967), 443–478.

7.6 CLELAND, R. "Extensibility of isolated cell walls: measurement and changes during cell elongation," *Planta*, 74 (1967), 197–209.

7.7 EVERT, R. E., L. MURMANIS, AND I. B. SACHS. "Another view of the ultrastructure of *Cucurbita* phloem," *Ann. Bot.* (N.S.), 120 (1966), 563–584.

7.8 ESAU, K. "Anatomy and cytology of *Vitis* phloem," *Hilgardia*, 37 (1965), 17–72.

7.9 Wooding, F. B. P. "Fine structure and development of phloem sieve tube content," *Protoplasma*, 64 (1967), 315–324.
7.10 Canny, M. J. "The mechanism of translocation," *Ann. Bot.* (N.S.), 104 (1962), 603–617.
7.11 Zimmerman, M. H. "Absorption and translocation: transport in the phloem," *Ann. Rev. Plant Phys.*, 11 (1960), 167–190.
7.12 Zimmerman, M. H. "Movement of organic substances in trees," *Science*, 133 (1961), 73–79.
7.13 Esau, K. "Development and structure of the phloem tissue, II," *Bot. Rev.*, (1950), 67–114.
7.14 Eschrich, W. "Kallose," *Protoplasma*, 47 (1956), 487–530.
7.15 Currier, H. B., K. Esau, and V. I. Cheadle. "Plasmolytic studies of phloem," *Amer. J. Bot.*, 42 (1955), 68–81.
7.16 Esau, K. "Origin and development of primary vascular tissues in seed plants," *Bot. Rev.*, 9 (1943), 125–206.
7.17 Esau, K. "Primary vascular differentiation in plants," *Biol. Revs.*, 29 (1954), 46–86.
7.18 Sinnott, E. W., and R. Bloch. "The cytoplasmic basis of intercellular patterns in vascular differentiation," *Amer. J. Bot.*, 32 (1945), 151–156.
7.19 Jacobs, W. P. "The role of auxin in differentiation of xylem around a wound," *Amer. J. Bot.*, 39 (1952), 301–309.
7.20 Fosket, D. E., and L. W. Roberts. "Induction of wound-vessel differentiation in isolated *Coleus* stems in vitro," *Amer. J. Bot.*, 51 (1964), 19–25.
7.21 Thompson, N. P. "The time course of sieve tube and xylem cell regeneration and their anatomical orientation in *Coleus* stems," *Amer. J. Bot.*, 54 (1967), 588–595.
7.22 Thompson, N. P., and W. P. Jacobs. "Polarity of IAA: effect on sieve-tube and xylem regeneration in *Coleus* and tomato stems, *Plant Physiol*, 41 (1966), 673–682.
7.23 Clutter, M. E. "Hormonal induction of vascular tissue in tobacco pith *in vitro*," *Science*, 132 (1960), 548–549.
7.24 Torrey, J. G. "The initiation of organized development in plants," in *Advances in Morphogenesis*, 5 (1966), 39–91.
7.25 Wetmore, R. H., and J. P. Rier. "Experimental induction of vascular tissues in callus of angiosperms," *Amer. J. Bot.*, 50 (1963), 418–430.
7.26 Wetmore, R. H., A. E. DeMaggio, and J. P. Rier. "Contemporary outlook on the differentiation of vascular tissues," *Phytomorphology*, 14 (1964), 203–217.

123 The flower of ***Helleborus niger*** (Christmas rose) showing perianth, numerous anthers, and three styles. × 2.

8

The Reproductive Tissues

FOR MANY of us, flowers, with their fascinating and often beautiful forms, colors, and scents, constitute our first introduction to plants. Flowers are, of course, the structures which contain the male and female sex organs (*stamens* and *carpels*). The elegant forms, the beautiful colors, and powerful scents appear to be the products of an evolution that has encompassed both plants and animals, especially insects. The complex forms and patterns of pigmentation, especially in the *perianth* parts (the *sepals* and *petals*), allow appropriate insects to recognize flowers as a source of food (pollen and nectar) in daylight, and the scents (not all of which are pleasant to us) appear to serve a similar function both by day and by night.

Figure 123 shows the arrangement of the floral organs in a simple flower, that of *Helleborus*. This species has numerous stamens, each composed of a thin stalk, the *filament*, and an *anther*, within which the pollen is formed. Three *styles*, prolongations of the upper part of the three separate carpels, appear in the center of the flower. At the tips of the styles are small *stigmatic discs*, upon which pollen becomes trapped and germinates. In many species, male and female sex organs are not formed in every flower. An example is shown in Figure 124, a transverse section of the compound inflorescence of *Solidago*. This particular inflorescence has twenty-one individual *florets*, of which five are sterile, four are female, and twelve bisexual (the leaf-like structures with deeply stained wings are *bracts* or floral leaves). A section through a bisexual floret and part of a female one is shown in more detail in Figure 125. In both cases the section passes through the centrally placed styles at the level of the stigmatic tissue (very densely stained), whereas in the bisexual floret one can distinguish the five anthers, each with four anther sacs, within which lie partly differentiated pollen grains.

Flowers are produced at shoot apices and the floral organs develop on the floral apex in a manner somewhat similar to that of leaves at the vegetative apex. This fact has led many workers to call floral organs "modified leaves." We feel that this terminology and approach are unfortunate and misleading. You have seen in earlier chapters that the character of the structures produced by any one vegetative shoot apex during the course of a growing season can range from bud scales to mature foliage leaves (Chapter 4). Recall again the difference between the aerial and aquatic leaves of water plants, a difference inducible simply by changing the water level in the tank in which the plants are growing. These examples indicate very clearly that the exact form of the structures that develop at the shoot apex is controlled by a variety of factors. In the sense that any structure which forms at the apex is the product of a particular sequence of growth and cellular differentiation, all shoot structures must have certain features in common. Since, in detail, no two successive leaves produced by the one shoot apex are ever identical (and indeed they may be very different if the conditions of growth at the apex have altered), we do not see that it is useful to regard either bud scales and other cataphylls or floral organs as "modified leaves." This is especially so when one considers those cases in which floral organs are not photosynthetic.

Although the developmental anatomy of floral induction is well studied in a variety of species (*8.1*), the really basic questions are still unanswered (*8.2*). In those plants in which flowering can be induced by exposing them to long nights or short nights (*8.3*), intensive physiological study has shown that some stimulus is formed in the leaves which leads to a change in the pattern of growth and differentiation of the vegetative apex. The length of time for which this stimulus must act varies in different plants. In one of the most studied long-night plants (a variety of *Xanthium*) a single long night is sufficient to induce the changes that lead to flower initiation. In most cases, at least several long nights are required and in some plants the stimulus must be maintained for several weeks. In these cases, if the plant is returned to an unfavourable photoperiod, further development of the apex stops, even though flower primordia have been induced.

The nature of this primary stimulus, its initial effect upon the vegetative apex, and indeed, the identity of the cells that receive the stimulus, are all major problems for future research.

The Stamen and Pollen Production

Figure 126 shows the filament, anther sacs, and the *connective* between them of a stamen of *Crocus*. Each sac has just opened by the formation of a longitudinal slit and the pollen grains, seen

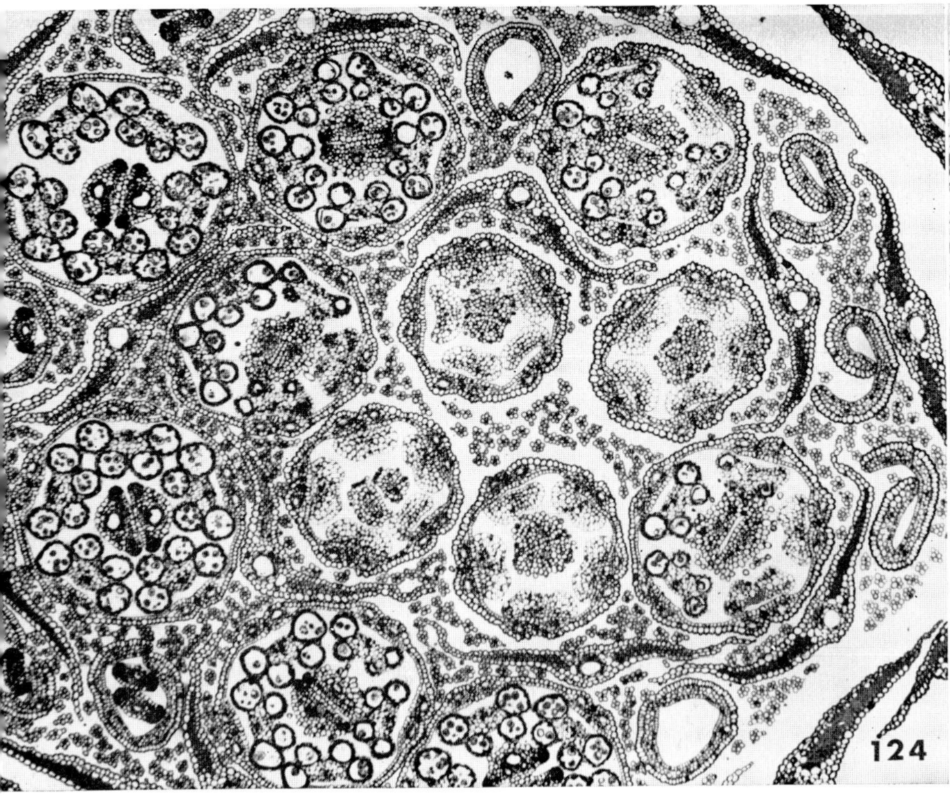

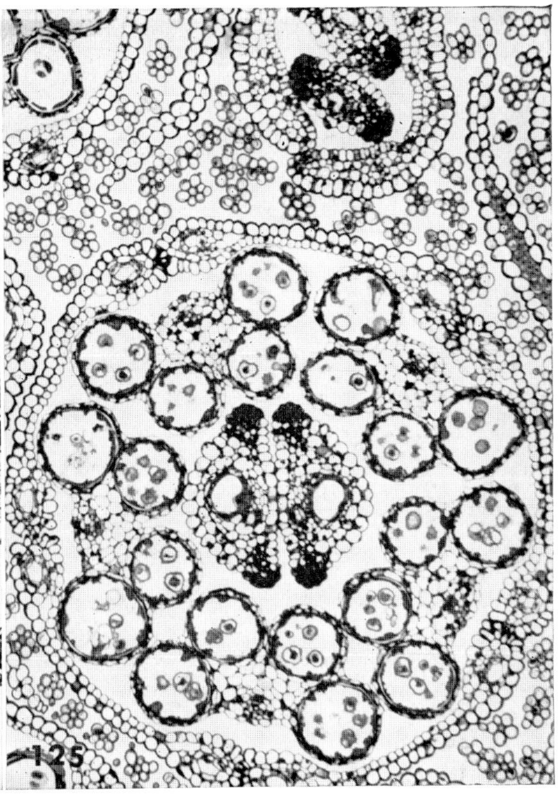

124 **Transverse section of the composite head of golden rod (*Solidago* spp.), showing a number of florets and bracts. PAS/toluidine blue. × 84.**

125 **Transverse section of a bisexual floret from Figure 124. Note the five fused petals and five anthers, each with four pollen sacs containing a few immature pollen grains. The section has passed through the styles at the level of the stigmatic discs (densely stained). PAS/toluidine blue. × 210.**

at higher magnification in Figure 127, are being released. The shape of the pollen grain, the position and number of its pores and the ornamentation of the pores and outer surface (Figures 128 and 129) are often species-specific characters (*8.4*). The outer layer of the wall is impregnated with sporopollenin, a polymer of unknown chemical structure which resists attack by the organisms in the soil. Much of this diagnostic pattern of the pollen grains therefore survives in fossil pollen and is of immense value in studying the history of the vegetation in different areas.

Each mature pollen grain is a product of *meiosis*, the reduction division in which the diploid number of chromosomes is reduced to the haploid number characteristic of the *gametes*. Some stages in the formation of pollen grains are shown in Figures 130–140.

The *primary sporogenous tissue* and its ensheathing layer, the *tapetum*, are differentiated quite early in the life of a stamen. It is generally believed that the tapetal cells act as a "nurse tissue," supplying the *microsporocytes* with essential nutrients during their meiotic divisions (but see *8.5*). In the case of *Solidago* illustrated here, the microsporocytes themselves appear to synthesize a gelatinous material that accumulates in large vacuoles and surrounds the nucleus (Figure 130, small arrows). Material with similar staining properties appears at the cell margins, and the microsporocytes come to be encased within it (Figure 132) (*8.6, 8.7*). Figure 131 shows a Feulgen stain of cells in the same stage as that shown in Figure 130, and it is quite clear that the microsporocytes have entered prophase of meiosis at the time that this secretion becomes evident.

Towards the end of prophase (Figure 132), the microsporocytes still appear to be connected to one another through this thick external layer by large cytoplasmic bridges (Figure 132, arrows), but these bridges are not evident at metaphase or in later stages of the meiotic divisions (Figures 133–138). At the end of meiosis, each microsporocyte consists of a mass of cytoplasm with four aggre-

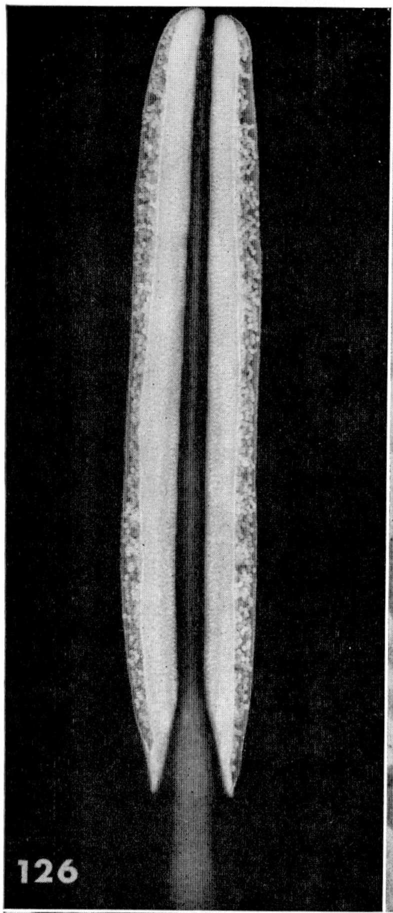

126

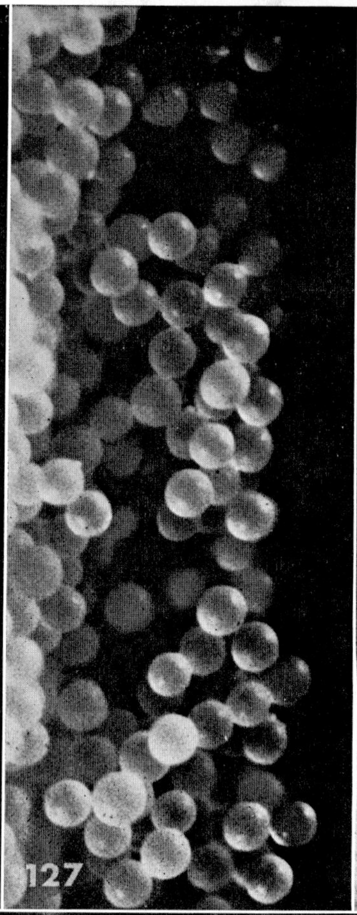

127

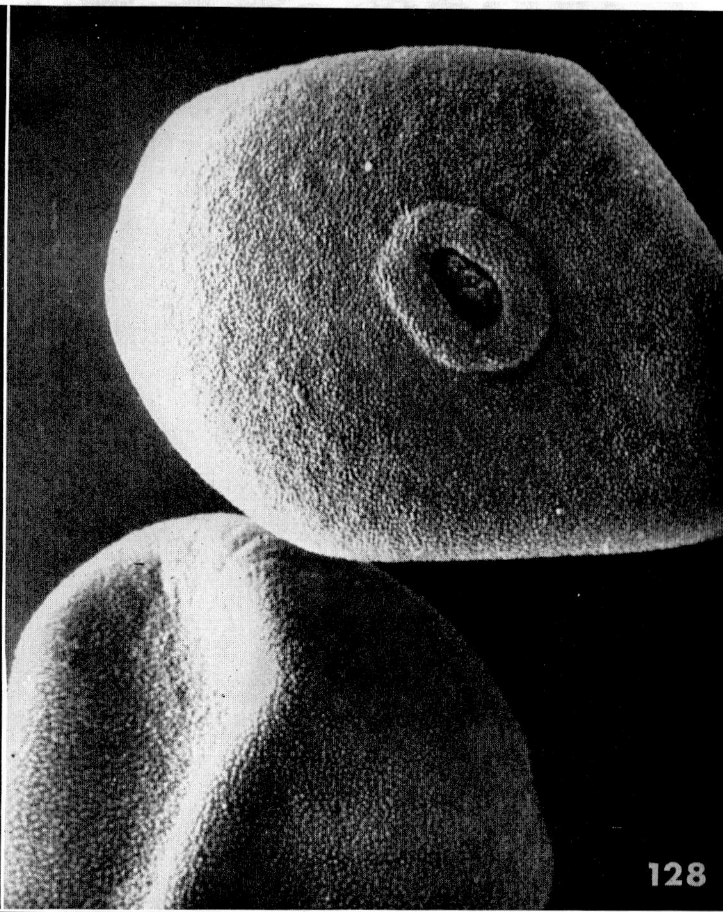

128

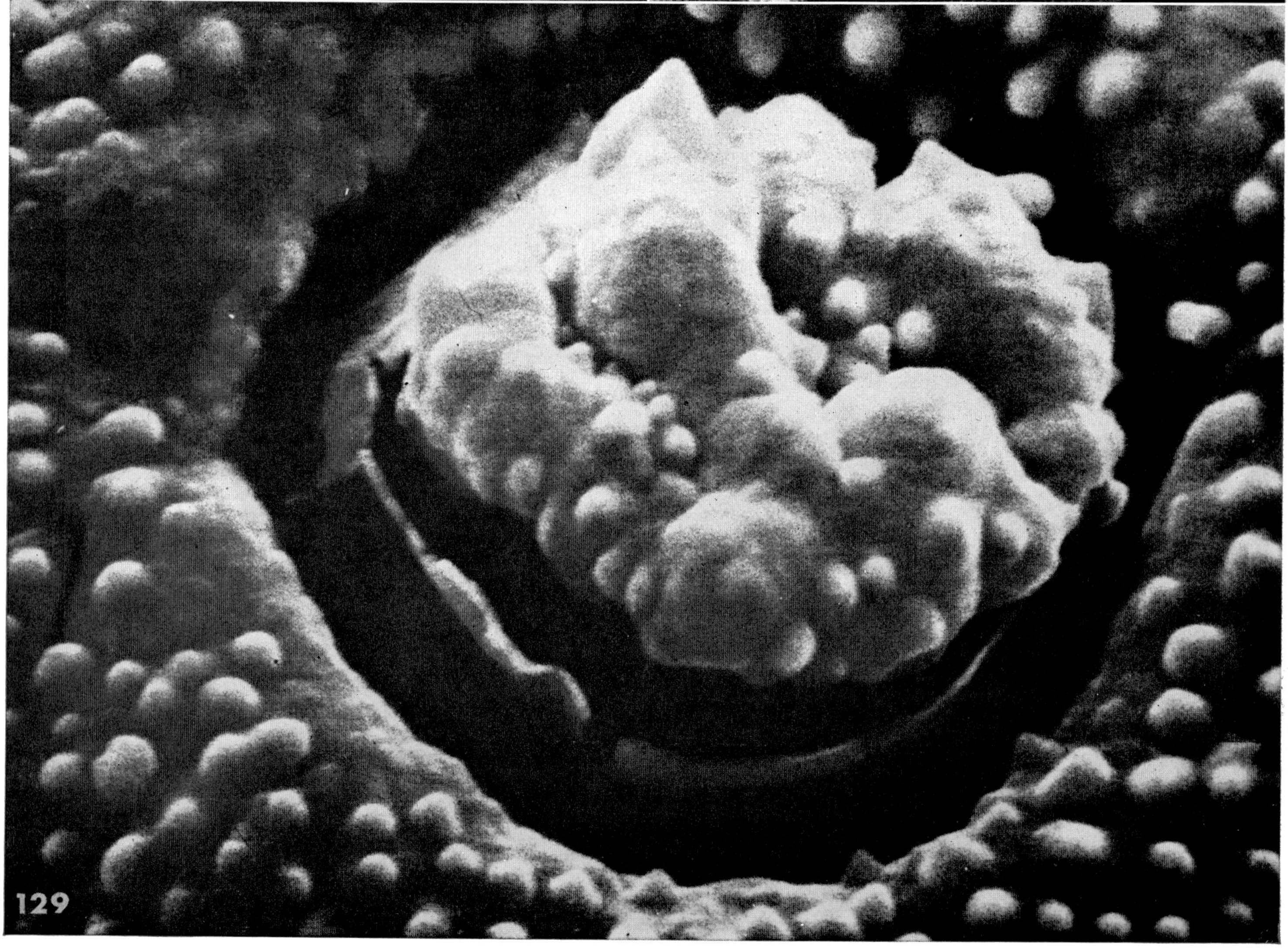

129

126 Part of a stamen from the flower of *Crocus* spp., showing the dehiscence of the anther and the release of the pollen grains. × 4.4.

127 *Crocus* pollen at higher magnification. × 56.

128 Electron micrograph (Cambridge Stereoscan) of pollen from *Triticum spelta*. This special type of electron microscope gives a three dimensional image of surfaces, and shows the texture of the pollen grain coat. Note the pore with its raised margin. The pollen tube emerges from this pore when the pollen grain germinates. × 1360.

129 Detail of a pore similar to that shown in Figure 128. Note the elaborate pattern of the wall over the pore. × 33,000. (Figures 128 and 129 courtesy Mr. J. R. Pilcher.)

gates of chromosomes, one each at the corners of a tetrahedron (Figure 137). These masses of chromosomes gradually disperse to form four nuclei, and the cytoplasm cleaves to form a *tetrad* of *microspores,* each with its own nucleus. In the normal course of events, each of these microspores would form its own wall (*8.8*) and be released as an individual pollen grain. In this species, microspores begin to degenerate at this stage, and development does not proceed beyond the earlier stages of wall production (Figure 138, arrows). Sections through a more mature anther which show the normal later stages are illustrated for *Lilium* in Figures 139 and 140.

Although the behavior of chromosomes in meiosis has been studied since the turn of the century, in general the fixatives used have not preserved the cytoplasmic contents of the microsporocytes. Furthermore, very little information has been published on the fine structure of meiotic cells in higher plants (*8.9*). Many of the photomicrographs reproduced here contain structures (other than the chromosomes) whose significance is not yet known (Figures 133 and 134). Since the

On the following page

130-138 Stages of meiosis and early stages of pollen grain formation in the anthers of *Solidago* spp. In most cases the sections include part of the tapetal layer (T), which surrounds the area in which the microsporocytes develop. All, × 1250.

130 Early prophase of meiosis showing the nuclei (N) surrounded by vacuoles (small arrows) rich in material similar to that which is accumulating at the cell surface (large arrow). PAS/toluidine blue.

131 Early prophase chromosomes, seen after Feulgen staining.

132 Late prophase. The individual cells are surrounded by extracellular material derived presumably from the content of the vacuoles in Figure 130. Note the numerous cytoplasmic bridges (arrows) between the microsporocytes. N, nucleus. PAS/toluidine blue.

133 and **134** Meiotic chromosomes (Ch), spindle fibers (SF) and various unknown structures (arrows, Figure 134). Each microspore is now surrounded by a thick wall (CW). PAS/toluidine blue.

135 and **136** Feulgen-stained meiotic chromosomes (Ch).

137 Early tetrad formation. The inset shows part of the chromosomes of each of the four microspores. PAS/toluidine blue.

138 Tetrads of microspores, showing the beginning of differentiation of the wall of the pollen grain (small arrows). Note the degeneration of some microspores. PAS/toluidine blue.

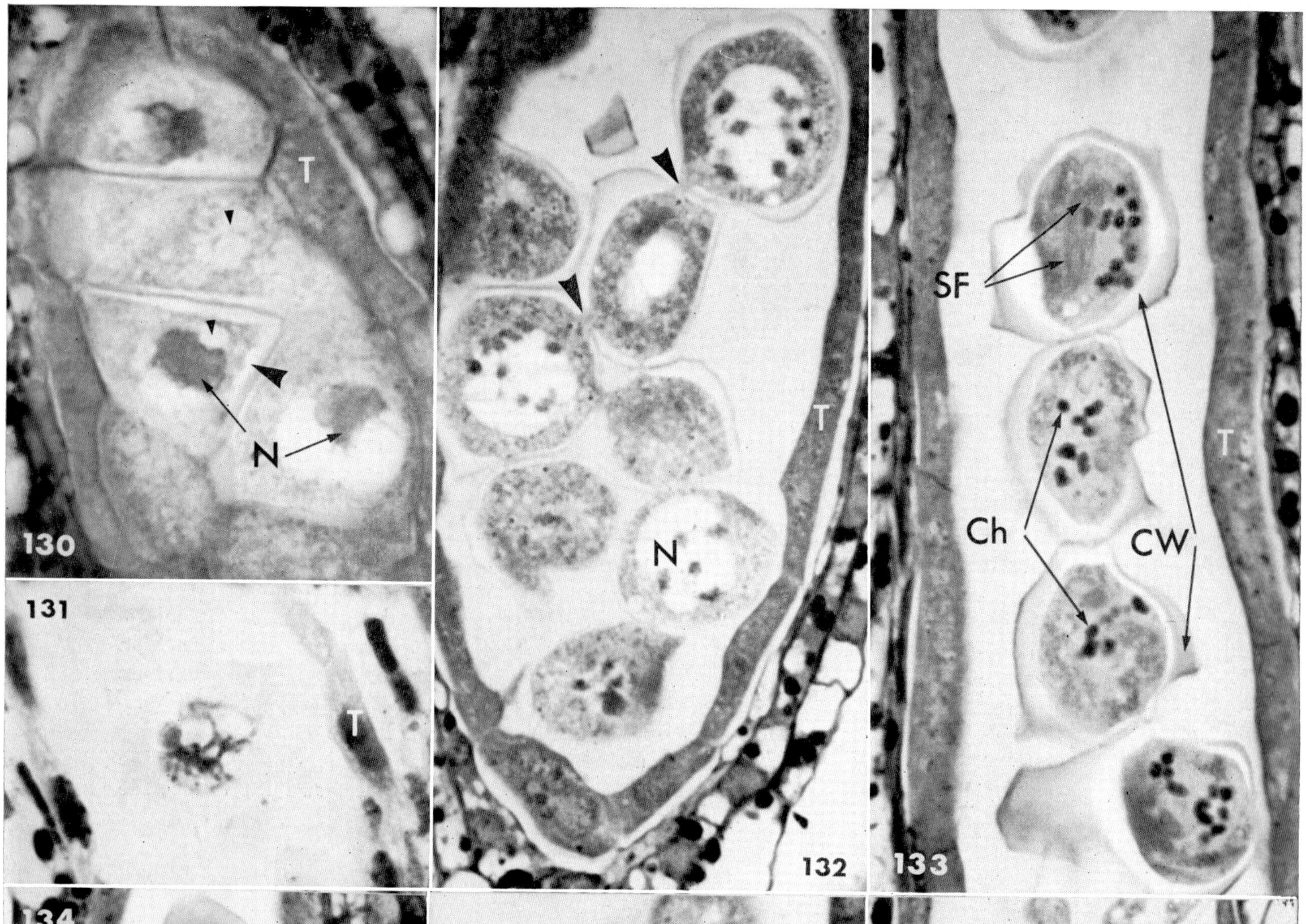
130
T
N
131
T
132
T
N
133
SF
Ch
CW
T
134
T
135
Ch
T
136
Ch
137
T
138
T

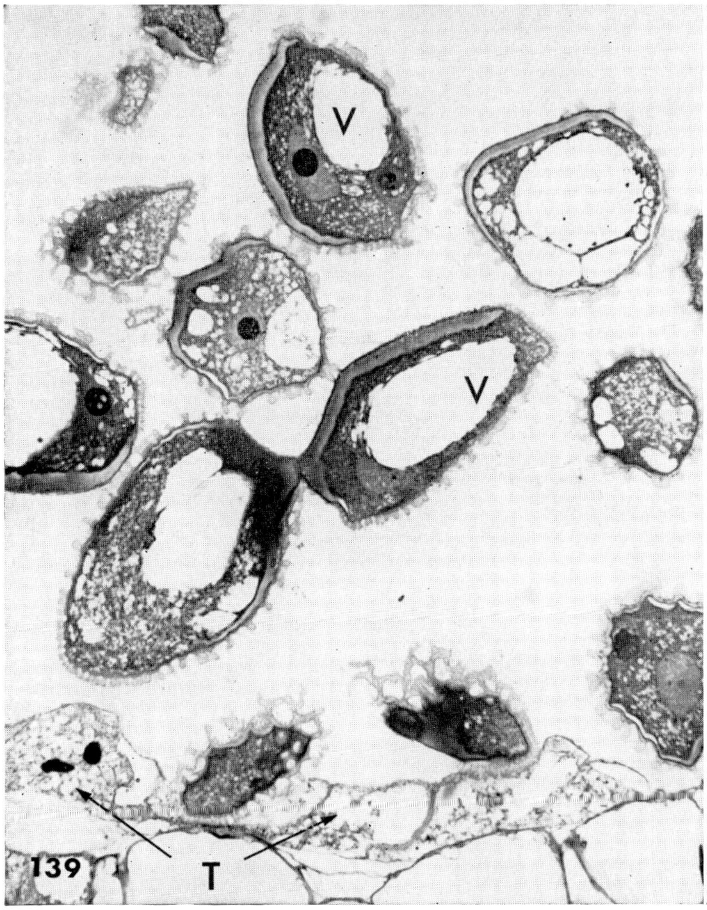

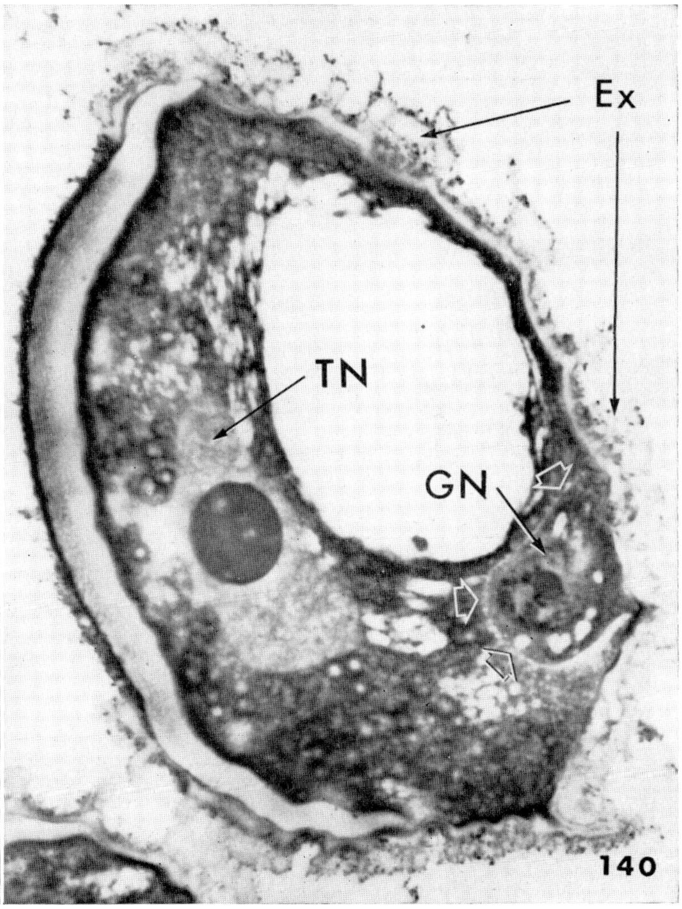

139 and 140 Transverse section through part of a lily anther showing maturing pollen. The intricate ornamentation of the pollen exine (Ex) has developed and the cells of the anther tapetum (T) are degenerating (Figure 139). The exine pattern is not uniform over the surface and an unornamented furrow of quite different staining properties is developed in each grain (Figures 139 and 140). A mature pollen grain contains a tube nucleus (TN) and a generative nucleus (GN), each of quite different morphology (Figure 140). The generative nucleus is enclosed in a small cell that is separated from the rest of the contents of the pollen grain by a thin wall, unstained either by toluidine blue or by the PAS procedure (arrows, Figure 140). V, vacuole. PAS/toluidine blue. Figure 139, × 500; Figure 140, × 1625.

microspores degenerate at the end of meiosis in *Solidago,* (the plant used for the illustrations), the behavior of the chromosomes in this plant may well be irregular, and you are referred to the general references for a description of the chromosomal events in meiosis.

When a pollen grain germinates on the stigmatic surface of the style, a *pollen tube* emerges from one of the pores and grows down through the stylar tissues towards the *ovules.* In some species, the ovular tissues appear to exert a chemotropic influence upon the growing tubes, guiding them towards the ovules (*8.10, 8.11, 8.12*). In those plants in which the pollen tubes must grow a considerable distance (for example, maize, in which pollen tubes reach a length of twenty to thirty centimeters before reaching the ovule), it is believed that the pollen tubes receive some nutrition from the stylar tissues en route.

The Ovule and Megasporogenesis

The ovules, which may be solitary or more than several hundred in number, arise from the tissues of the *placenta* (Figure 141). Each ovule consists of a stalk, the *funiculus,* which bears the *nucellus,* surrounded by either one or two *integuments* (Figures 143 and 144). During ovule development, growth may be unequal, leading to various degrees of curvature of the body of the ovule. One of the

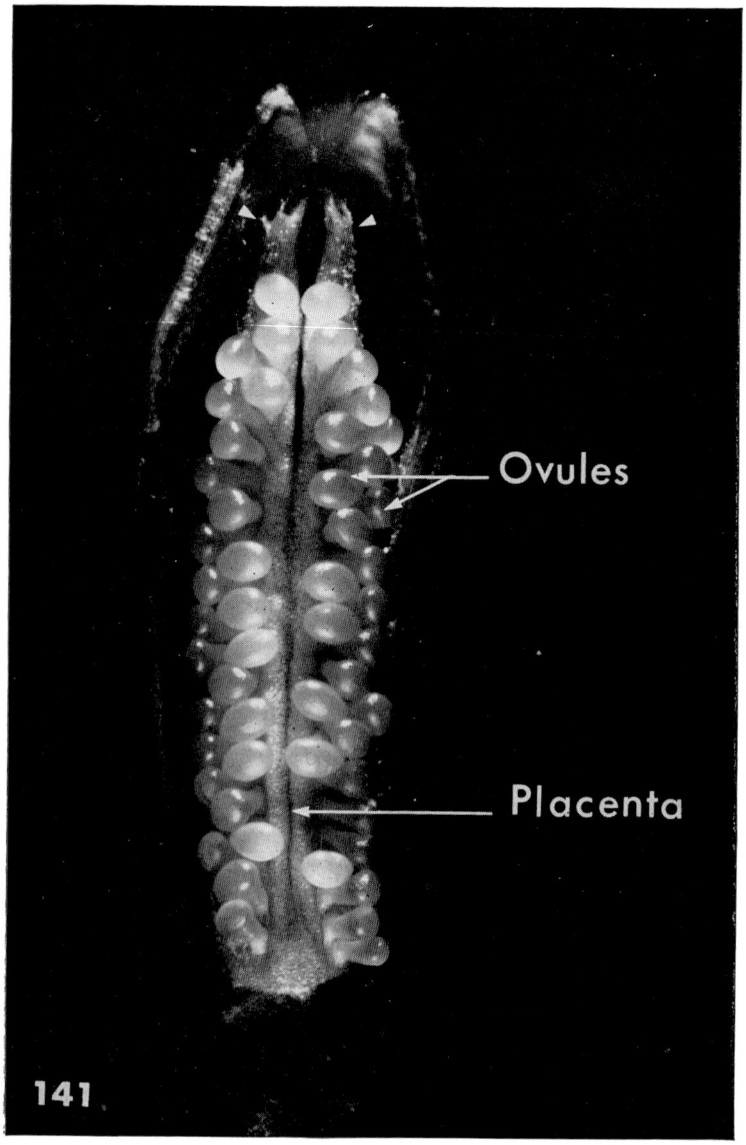

141 **Dissected ovary of *Dianthus* spp. (carnation), showing numerous ovules attached to the placentae. The placentae are connected to the top of the ovary (small arrows) so that, in contrast to the statements expressed in many texts, the placentation in this genus is *not* free central. × 14. (Specimen kindly prepared by Mrs. D. J. Carr.)**

most common types, designated *anatropous*, arises in this way when the *micropyle* (Figure 143, asterisk) is directed towards the base of the funiculus (Figures 143 and 144).

Commonly a single hypodermal cell of the nucellus develops into the *megaspore mother cell* which undergoes meiosis to form a tetrad of *megaspores*, each with its own haploid nucleus contained within a definite cell. Figure 145 shows a section through a megaspore mother cell at the first anaphase of meiosis. In most cases, only one of these megaspores goes on to form the *embryo sac*. However in quite a number of genera, the haploid nuclei formed by meiosis are not completely separated into cells and more than one of these nuclei may subsequently become involved in embryo sac formation.

In most cases, one of the nuclei formed by meiosis from the megaspore mother cell becomes enclosed within a cell membrane and forms an *egg*. The egg nucleus fuses with one of the *gametic nuclei* released from the pollen tube. This restores the diploid chromosome number in the *zygote*, the product of this fusion of haploid nuclei. In a great many angiosperms, the other gametic nucleus released from the pollen tube fuses with two or more of the other nuclei of the embryo sac to form an *endosperm nucleus*. The growth and further division of this nucleus gives rise to the *endosperm*, a nutritive tissue which, when formed in this way, is always at least *triploid* (high levels of ploidy have been recorded for the nuclei of mature endosperm in some species). Subsequent to this act of *double fertilization*, a new chapter of events unfold —the development of the zygote into the embryo and the formation of the seed.

While information is meager in general about the physiology of sex and reproduction in higher plants, our ignorance is deepest in matters con-

142 **Longitudinal section through a floret of *Solidago* showing the relationship of the ovule (O) within the inferior ovary to the anthers (A). The base of the style is shown by an asterisk. The section passes through a laticifer (L). PAS/toluidine blue. × 210.**

143 **Longitudinal section of an ovule of *Solidago*, showing the integument (In), micropyle (asterisk), and funiculus (F). PAS/toluidine blue. × 450.**

144 **The nucellus (Nuc) and megaspore mother cell (MMC) in an ovule of *Solidago*. PAS/toluidine blue. × 450.**

145 **Higher magnification of the megaspore mother cell of *Solidago* shown in Figure 144. This cell is in the anaphase stage of meiosis. Ch, chromosomes. Toluidine blue. × 2100.**

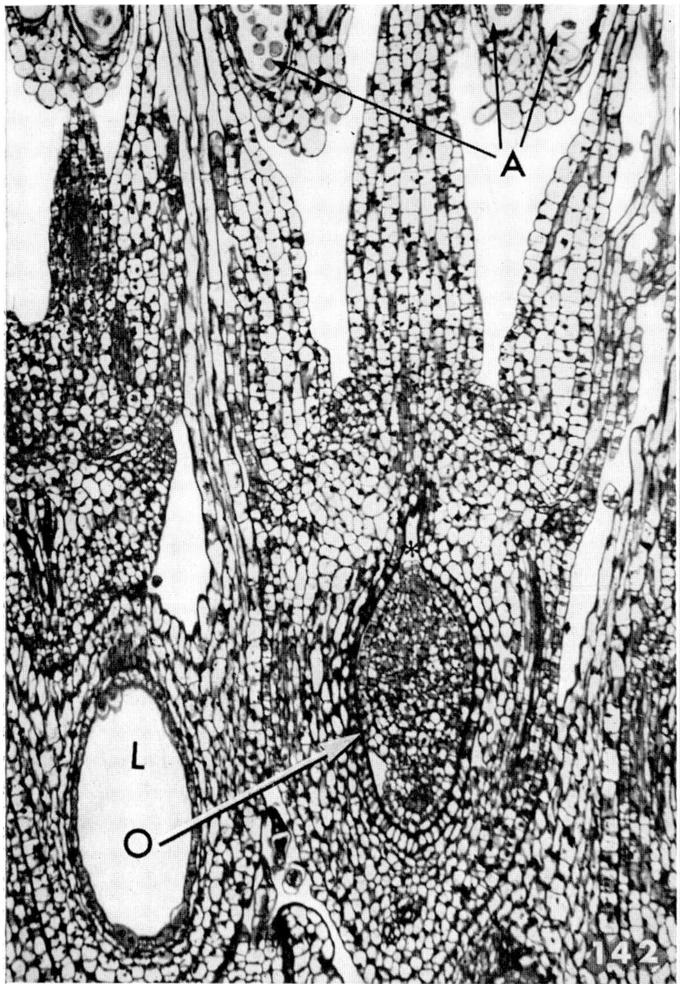
A
L
O
142

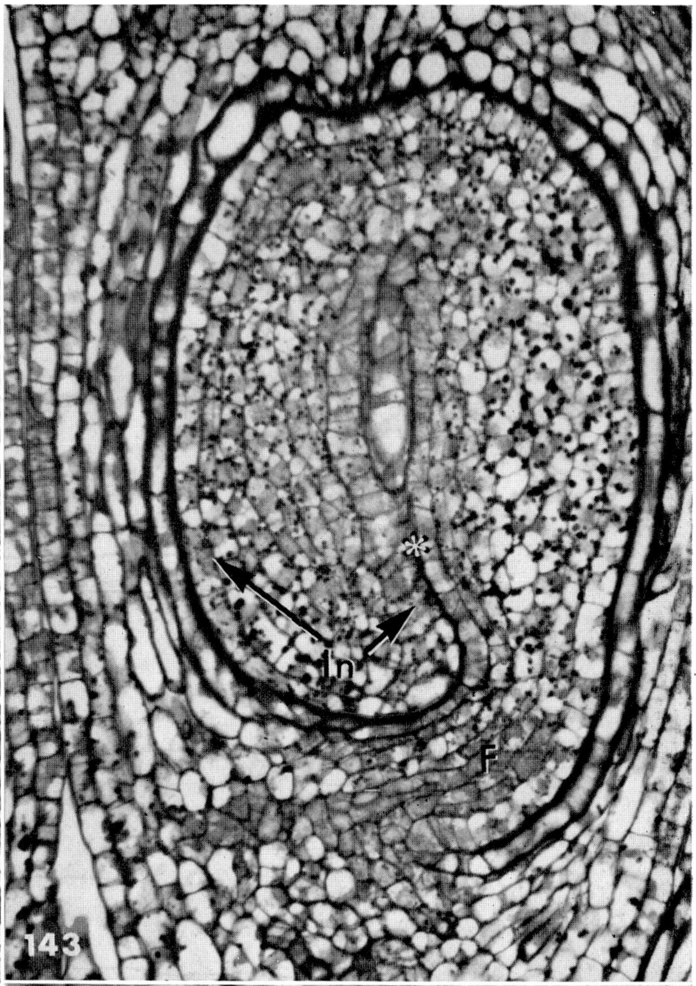
In
F
143

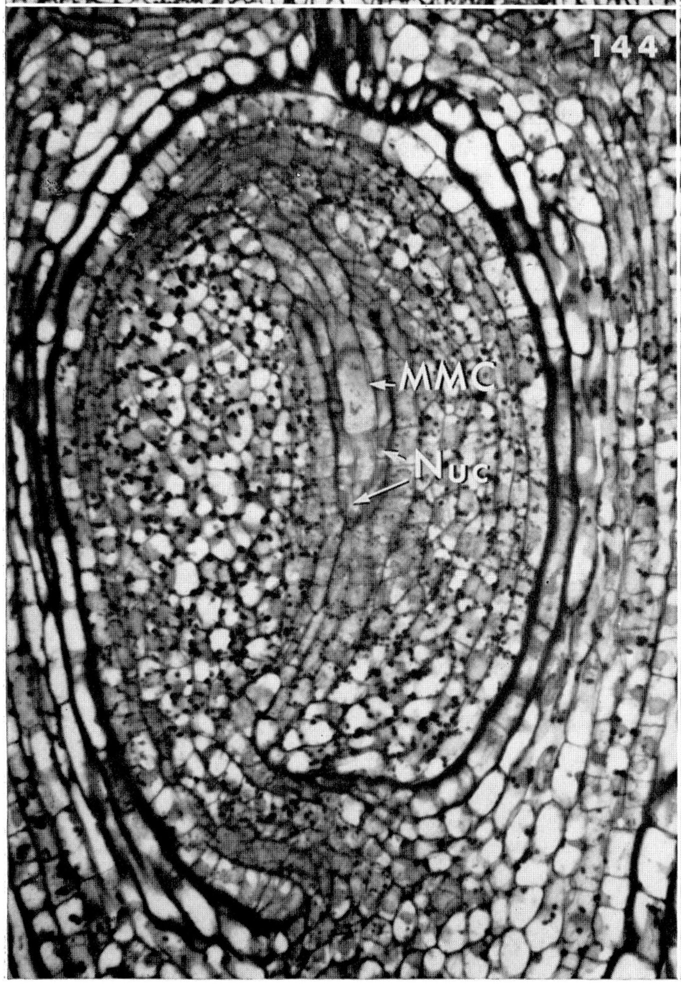
144
MMC
Nuc

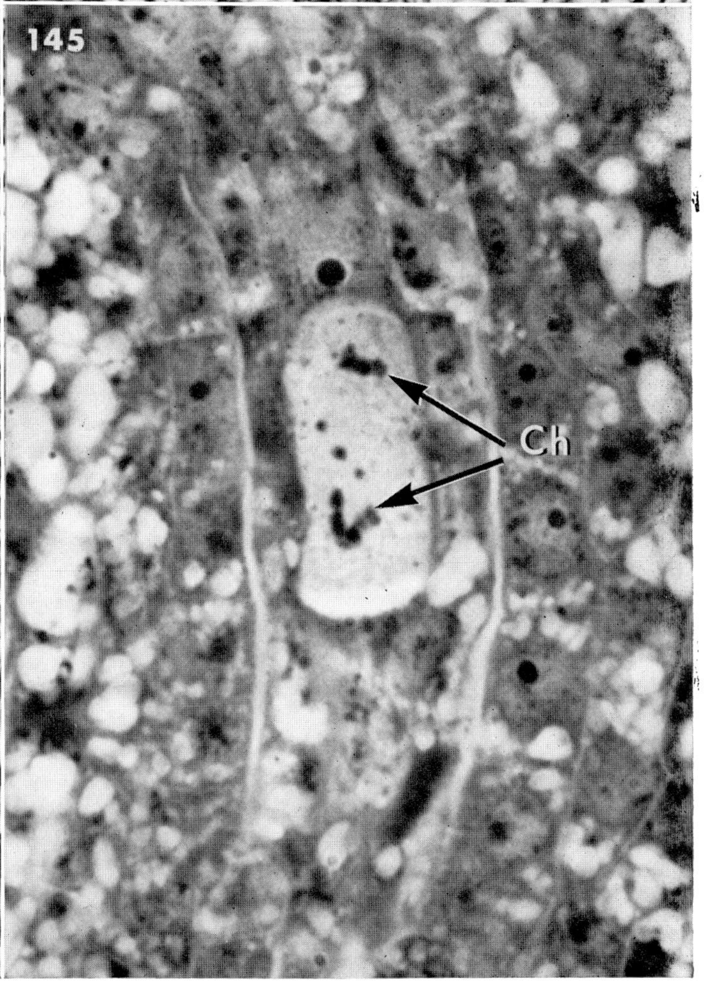
145
Ch

cerned with the development of the embryo sac and its behavior during and shortly after fertilization (*8.2, 8.13*). Unfortunately, these structures are often very difficult to preserve for electron microscopy. Until a great deal more is known of the fine structure of embryo sacs (*8.14, 8.15, 8.16, 8.17*), it will be impossible to pass beyond the limits of our present knowledge, or to understand the significance of the many diverse patterns of development that are known in different genera.

GENERAL REFERENCES

CHURCH, A. H. *Types of Floral Mechanism.* Oxford: Clarendon Press, 1908.

FOSTER, A. S., AND E. M. GIFFORD. *Comparative Morphology of Vascular Plants.* San Francisco: W. H. Freeman and Co., 1959.

HILLMAN, W. S. *The Physiology of Flowering.* New York: Holt, Rinehart & Winston, 1963.

MAHESHWARI, P. *An Introduction to the Embryology of Angiosperms.* New York: McGraw-Hill Book Co., 1950.

PERCIVAL, M. S. *Floral Biology.* Oxford: Pergamon Press, Ltd., 1965.

SALISBURY, F. B. *The Flowering Process.* Oxford: Pergamon Press, Ltd., 1963.

WARDLAW, C. W. *Embryogenesis in Plants.* New York: John Wiley & Sons, 1955.

REVIEWS AND RESEARCH PAPERS

8.1 NITSCH, J. P. "Physiology of flower and fruit development." *Encyclopaedia of Plant Physiology,* XV, Pt. 1, ed. W. Ruhland. Berlin: Springer-Verlag, 1965.

8.2 LANG, A. "Flower and fruit formation," *Encyclopaedia of Plant Physiology,* XV, Pt. 1, ed. W. Ruhland. Berlin: Springer-Verlag, 1965.

8.3 ZEEVART, J. A. D. "Climatic control of reproductive development," in "*Environmental Control of Plant Growth,*" ed. L. T. Evans. New York: Academic Press, 1963.

8.4 ERDTMAN, G. *Pollen Morphology and Plant Taxonomy.* Stockholm: Almqvist and Wiksell, 1952.

8.5 TAKATS, S. T. "An attempt to detect utilization of DNA breakdown products from the tapetum for DNA synthesis in the microspores of *Lilium longiflorum,*" *Amer. J. Bot.,* 49 (1962), 748–758.

8.6 ESCHRICH, W. "Elektronenmikroskopische untersuchungen an Pollenmutter-zellen-callose. IV. Mitteilung über callose," *Protoplasma,* 55 (1962), 419–22.

8.7 HESLOP-HARRISON, J. "Cytoplasmic continuities during spore formation in flowering plants," *Endeavour,* 25, (1966), 65–72.

8.8 ROWLEY, J. R. "Ubisch body development in *Poa annua,*" *Grana palynologica,* 4 (1963), 25–36.

8.9 TAYLOR, J. H. "Meiosis," in *Encyclopaedia of Plant Physiology,* XVIII, ed. W. Ruhland. Berlin: Springer-Verlag, 1967.

8.10 LINSKENS, H. F. *Pollen Physiology and Fertilization.* Amsterdam: North-Holland Publishing Co., 1964.

8.11 WELK, M., W. F. MILLINGTON, AND W. G. ROSEN. "Chemotropic activity and the pathway of the pollen tube in lily," *Amer. J. Bot.,* 52 (1965), 774–781.

8.12 DASHEK, W. V., AND W. G. ROSEN. "Electon microscopical localization of chemical components in the growth zone of lily pollen tubes," *Protoplasma,* 59 (1966), 192–204.
8.13 LANG, A. "Auxins in flowering," *Encyclopaedia of Plant Physiology,* XIV, ed. W. Ruhland. Berlin: Springer-Verlag, 1961.
8.14 JENSEN, W. A. "The composition and ultrastructure of the nucellus in cotton," *J. Ultrastr. Res.,* 13 (1965), 112–128.
8.15 JENSEN, W. A. "The ultrastructure and composition of the egg and central cell of cotton," *Amer. J. Bot.,* 52 (1965), 781–797.
8.16 JENSEN, W. A. "Cotton embryogenesis: the zygote," *Planta,* 79 (1968), 346–366.
8.17 JENSEN, W. A., AND D. B. FISHER. "Cotton embryogenesis: the entrance and discharge of the pollen tube in the embryo sac," *Planta,* 78 (1968), 158–183.

146 The photosensitive seeds of lettuce (*Lactuca sativa* var. Grand Rapids). × 18.

9

The Seed

Most of us are introduced to seeds as small children and perhaps you can still remember the fascination with which you first watched the daily progress of a germinating seed. Seeds are remarkable structures (Figure 146) and it is likely that the abundance of angiosperms in the modern flora is due to the evolution of seeds. Although all fertile seeds contain both an embryo and a food reserve enclosed within a protective coat, the detailed structure of the seed varies widely among the different genera of flowering plants. The size and stage of development of the embryo, the amount, location, composition of the food reserves, and the ontogeny of the protective coats are all variable. Figures 147, 149, and 150 illustrate three rather different types of seed.

The seed of cress shown in Figure 147 is not quite mature but it contains a simple embryo, differentiated only into two cotyledons, a hypocotyl, and a shoot and root apex. The provascular tissue of the cotyledons and hypocotyl and the first cells of the quiescent center of the root (Figure 148, white arrow) are evident. The cells of the *suspensor,* to which the developing embryo was attached during the early stages of its growth, have degenerated (Figure 148, asterisk). Food reserves are present both in the endosperm and in the embryo, and the seed is protected by a thick integument, rich in starch and hemicellulose.

The mature lettuce seed (Figure 146) is actually a fruit but consists of little more than the tissues of the embryo. Figure 149 shows a section of such a seed that had been allowed to germinate for six hours. The cotyledons are large and provascular tissue is present both in the cotyledons and in the hypocotyl. Almost all of the food reserves lie within the embryo, for the endosperm consists only of a thin layer that lies between the embryo and the seed coat. In the ungerminated seed, this layer of endosperm is pressed tightly around the embryo and germination cannot proceed until the mechanical restraint of this layer is removed. The seed coat is derived from an integument and the ovary wall.

In the seeds of grasses, the embryo develops much further before the seed matures (Figures 150 and 151). The shoot apex produces several immature leaves and the whole of the shoot is enclosed by a special organ, the *coleoptile.* After germination, this organ is sensitive to light and to gravity and guides the shoot system to the soil surface and protects it en route. The embryo contains one or more roots embedded within the storage tissues and separated from them by a layer of extracellular material, rich in polysaccharide, which is produced by a layer of secretory cells at the surface of the root. This layer and its products (Figure 151) are similar to those which form over the surface of lateral roots in some plants (see Figure 50).

The grass embryo has only one cotyledon, the *scutellum.* The endosperm is abundant and rich in protein and starch but its outermost layer, the *aleurone layer,* is free of starch and contains special protein-rich bodies, the *aleurone granules.* The food reserve is therefore present both in the tissues of the endosperm and in the embryo but the starch is usually present only in the endosperm.

A special layer of cells, the *scutellar epithelium,* lies at the boundary of the scutellum and the endosperm. Prior to germination, these cells are columnar, closely appressed, and filled with cytoplasm; only the most minute vacuoles can be observed, and the cells are totally devoid of starch (Figures 152 and 153). Striking changes take place in this layer during germination. The epithelial cells elongate, become separated from one another at their tips and grow out into the endosperm; in this state they are often termed *haustorial cells.* Vacuoles become prominent (Figure 154) and the cells accumulate an abundance of starch. This starch is synthesized from soluble precursors absorbed from the endosperm where starch hydrolysis takes place (Figure 155).

The protein-rich cells of the scutellum also change their appearance after germination (*9.1*). Before germination, these cells are packed with small protein-rich droplets but after germination, these droplets coalesce to form quite large vacuoles and their content of stored protein gradually disappears (Figures 152 and 154).

Studies on isolated aleurone layers of certain varieties of barley have shown that the cells of this layer can produce the starch hydrolyzing enzyme α-amylase when treated with the hormone gibberellic acid (*9.2*). It is thought that in the intact germinating grain, the natural gibberellins are synthesized within the embryo, diffuse to the aleurone layers, and intiate the synthesis and release of α-amylase. It seems unlikely, however, that the aleurone layer is the only source of starch hydrolyzing enzymes, because starch hydrolysis often begins in the endosperm layer immediately adjacent to the scutellar epithelium (see Figure 155).

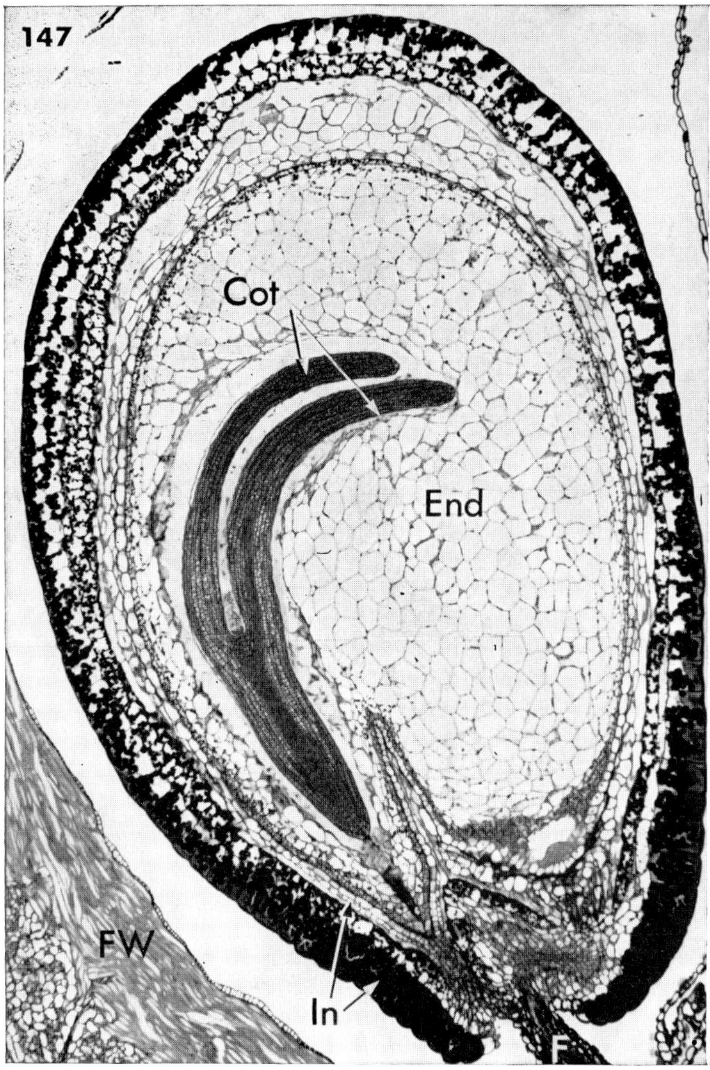

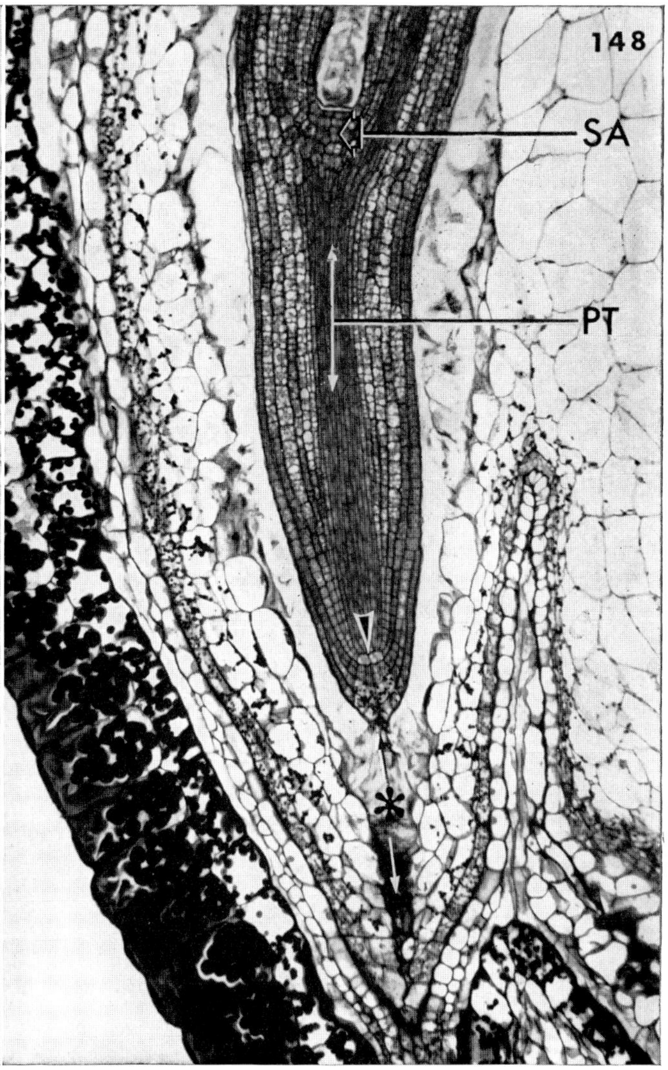

147 Longitudinal section of a seed of cress (*Lepidium* spp.), showing embryo with well developed cotyledons (Cot) and endosperm (End). The integuments (In), funiculus (F) and fruit wall (FW) can be seen. PAS/toluidine blue. × 84.

148 Detail of Figure 147. The suspensor (asterisk) has degenerated and the root apex appears to have developed a quiescent center (white arrow). Provascular tissue (PT) has differentiated in the embryo and the shoot apex (SA) has organized. Note the presence of starch (dark granules) and hemicellulose (thick, stained walls) in the integument. PAS/toluidine blue. × 210.

The seeds of a great many plants become dormant once the seed is mature (*9.3, 9.4, 9.5*). In the simplest cases, the dormancy is imposed by the impermeability of the seed coats, especially to water and to oxygen. Once this impermeable layer is removed or damaged (for example, by abrasion of the coat, treatment with strong acid, or by the attack of the soil microflora) germination proceeds immediately.

Somewhat more intriguing are the cases where germination is controlled by light. The light-promoted seeds of Grand Rapids lettuce (Figure 146) (*9.6*) and the light-inhibited seeds of *Phacelia tanacetifolia* (*9.7*) are among the better understood. In both of these cases, the embryo will germinate if dissected out from the seed, independent of the light regime to which it is exposed. Careful surgery of intact seeds has shown that it is the mechanical restraint of the endosperm in the region of the radicle which prevents germination. Just how the light stimulus leads to an alteration in the mechanical restraint imposed by the endosperm is not known but it is significant that gibberellins induce germination in both types of

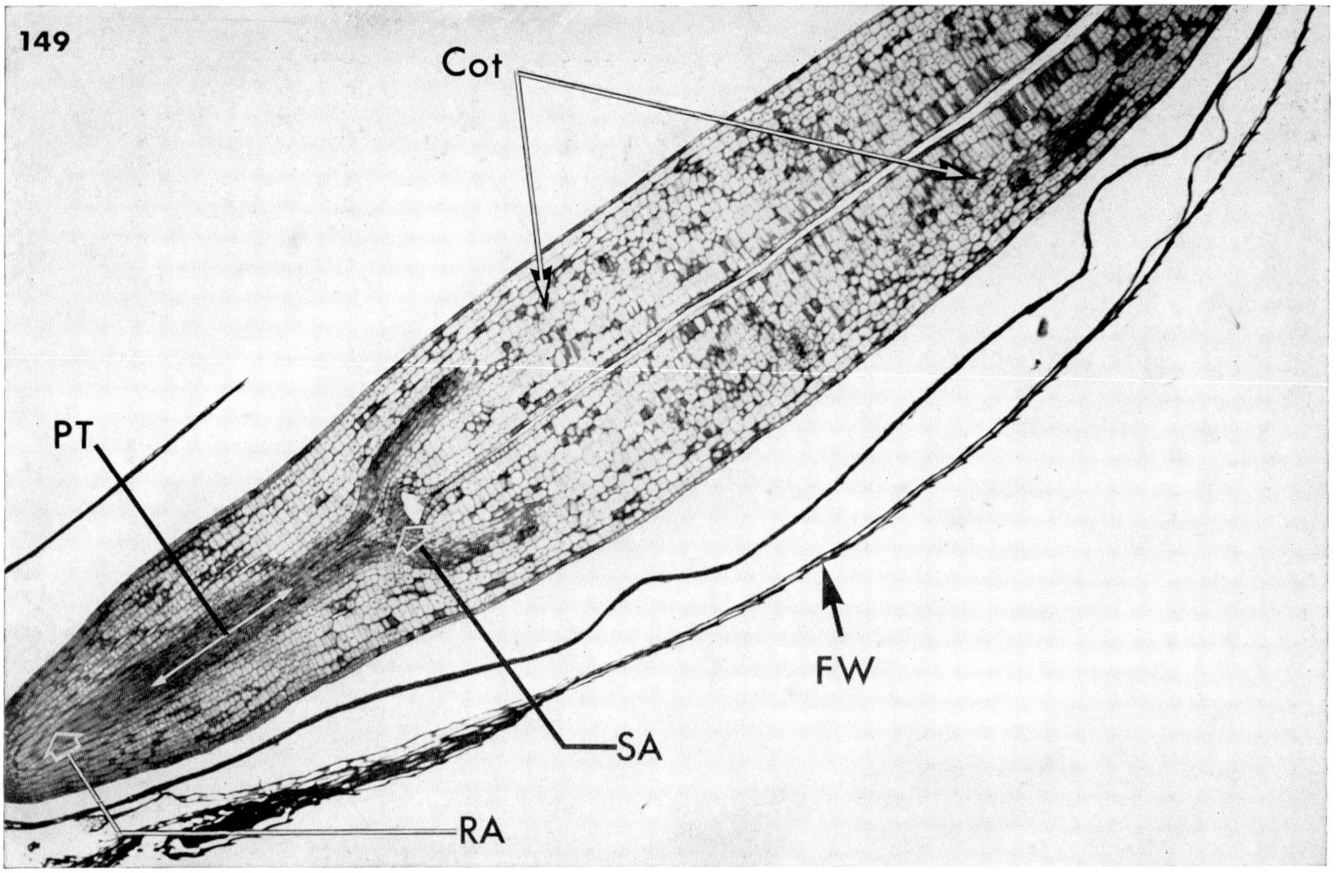

Captions for Figures 149 to 155 on page 106

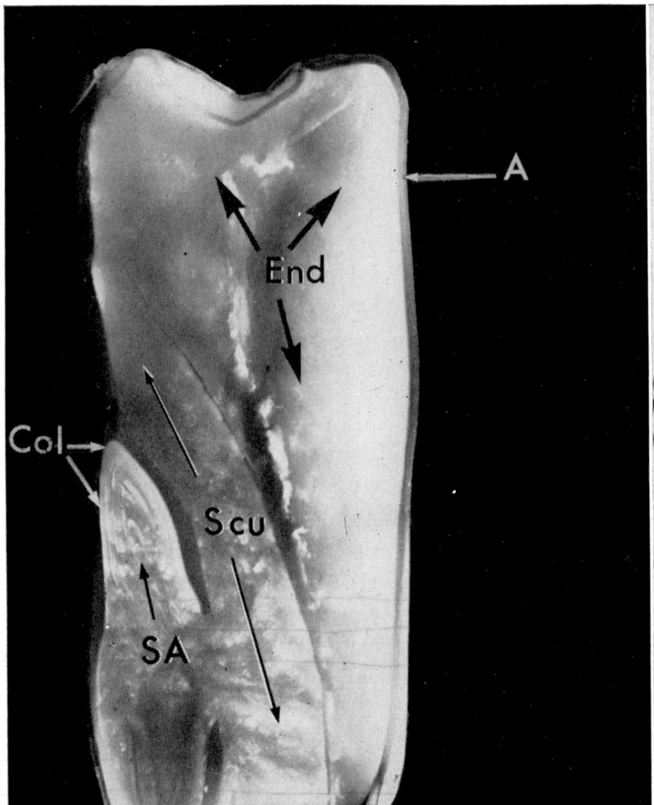

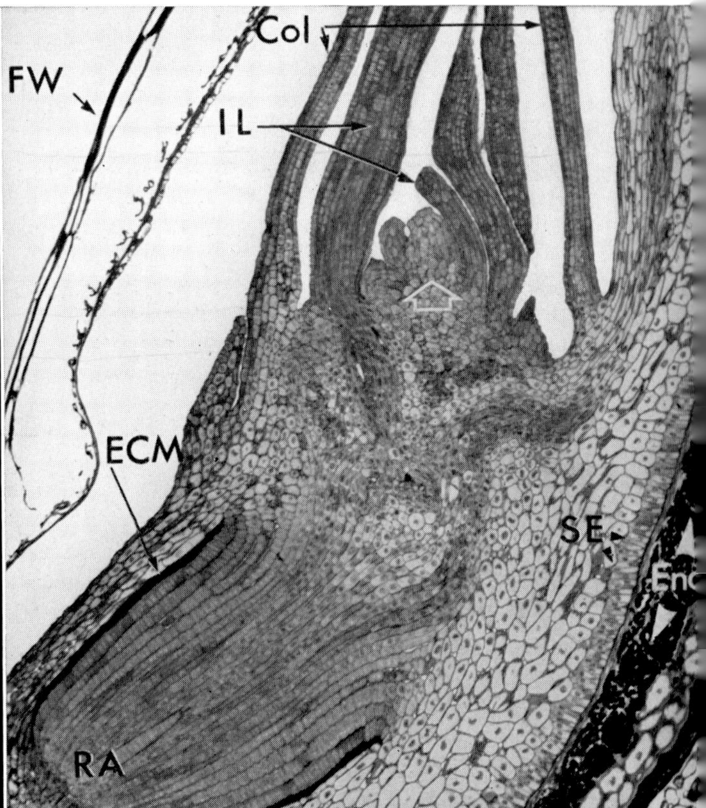

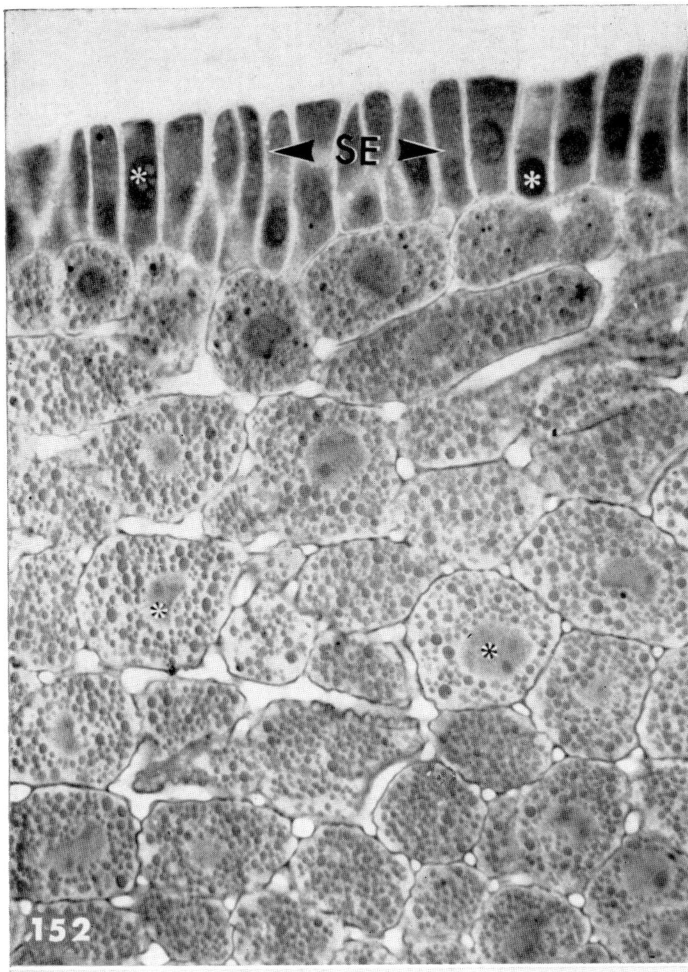
SE
152

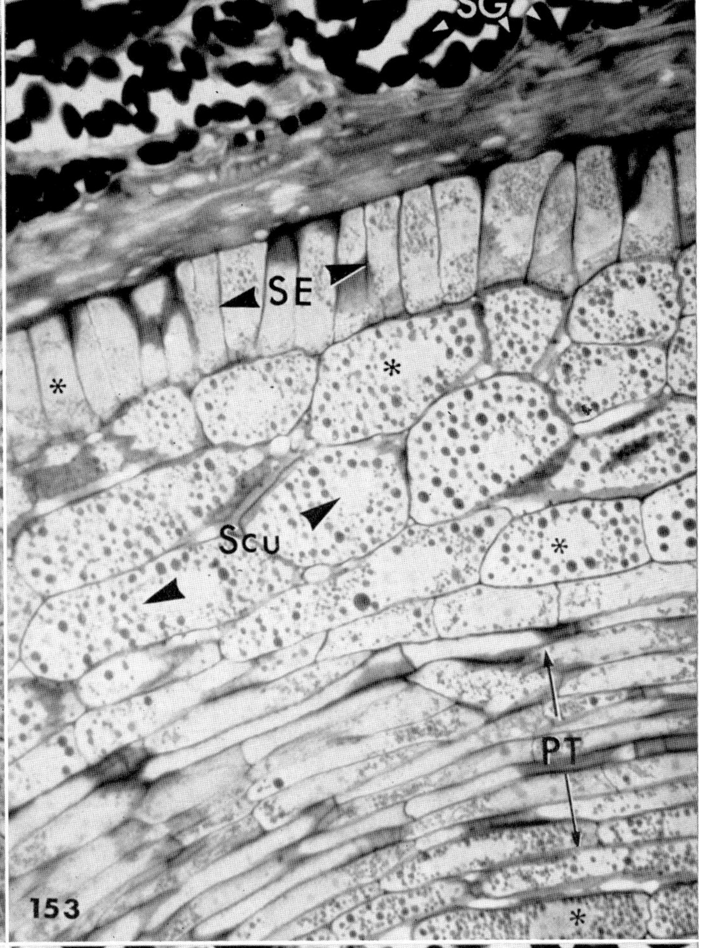
SG
SE
Scu
PT
153

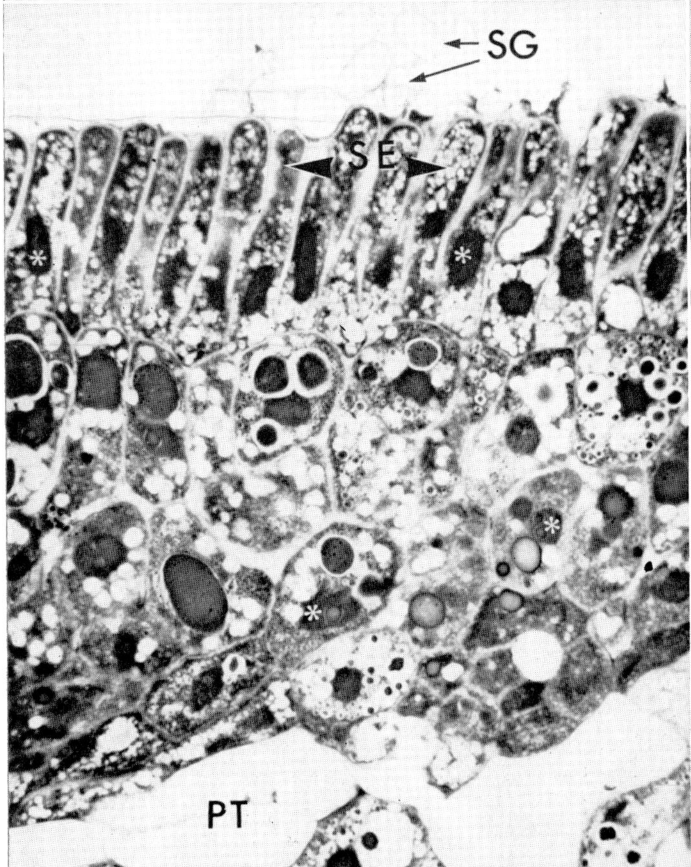
SG
SE
PT
154

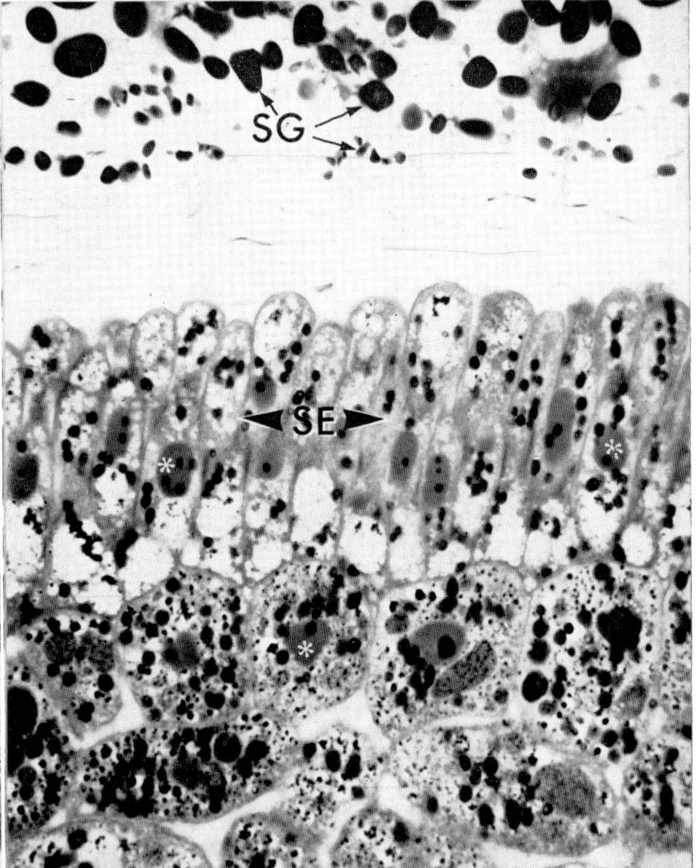
SG
SE
155

149 **Longitudinal section through a germinating seed of lettuce. The embryo has a well developed root apex (RA) and shoot apex (SA) and has differentiated provascular tissue (PT). Most of the food reserves are confined to the embryo, mainly in the large cotyledons (COT). FW, fruit wall. Toluidine blue. × 70.**

150 **Sagittal section of a soaked grain of maize showing the starchy endosperm (End) surrounded by the aleurone layer (A). The scutellum (Scu), coleoptile (Col), and the young root apex (RA) and shoot apex (SA) of the embryo can be seen. × 10.**

151 **Longitudinal section of the embryo and part of the starch-rich endosperm (End) of a soaked grain of wheat. Note the coleoptile (Col) which encloses a number of immature leaves (IL) that in turn subtend the shoot apex (open white arrow). The root apex (RA) of one of the seminal roots (see Figure 37) is evident and a thick layer of extracellular material (ECM) covers the root (compare this layer to that which surrounds the lateral roots of maize in Figure 50). The scutellar epithelium (SE) lies at the surface of the scutellum where it abuts the endosperm. FW, fruit wall. PAS/toluidine blue. × 80.**

152 and 153 **Section through the scutellum (Scu) of an ungerminated grain of wheat stained with acid fuchsin/toludine blue (Figure 152) and PAS/fast green (Figure 153). Nuclei are indicated by asterisks. Note the absence of starch (SG) from the scutellum and the arrangement of the storage protein droplets. SE, scutellar epithelium; PT, provascular tissue. × 550.**

154 and 155 **Sections through the scutellum of a grain of wheat germinated for about 40 hours, stained with acid fuchsin/toluidine blue (Figure 154) and PAS/fast green (Figure 155). Note the growth of the cells of the scutellar epithelium (SE), accumulation of starch (SG), and mobilization of storage protein. Starch has been mobilized from the endosperm near the tips of the haustorial cells (compare Figures 153 and 155). PT, provascular tissue. Nuclei are again indicated by asterisks. × 550.**

light-sensitive seeds, even when they are exposed to inhibitory light conditions.

Even more complex are the situations in which the embryo itself is dormant and will not germinate even if isolated from the seed coats. In many of these cases, there is a requirement for prolonged chilling, of a type normally received by the seeds as they lie beneath the snow for several months.

Dormancy is undoubtedly of great importance to the survival of angiosperms. In one of the simplest cases, hard seed coats (common in legumes) ensure that only a fraction of the total seed crop germinates in any one growing season, leaving a reservoir of ungerminated but still viable seed for later seasons. The more elaborate types of dormancy control, especially those based upon a chilling requirement, often ensure that the seeds germinate only in the spring when conditions are favorable for the establishment of the seedling.

GENERAL REFERENCES

CROCKER, W., AND L. V. BARTON. *Physiology of Seeds*, Boston, Mass.: Chronica Botanica Co., 1957.

ESAU, K. *Plant Anatomy*, 2d ed., Chaps. 19 and 20. New York: John Wiley & Sons, 1965.

HEINISCH, O. *Samenatlas*. Leipzig: Druckhaus Einheit, 1955.

MAYER, A. M., AND A. POLJAKOFF-MAYBER. *The Germination of Seeds*. Oxford: Pergamon Press, 1963.

NIETHAMMER, A., AND N. TIETZ. *Samen and Früchte*. The Hague: W. Junk, 1961.

Seeds, The Yearbook of Agriculture. Washington, D. C.: Department of Agriculture, 1961.

REVIEWS AND RESEARCH PAPERS

9.1 O'Brien, J. A. "Plastid development in the scutellum of *Triticum vulgare* and *Secale cereale*," *Amer. J. Bot.*, 38 (1951), 684–696.
9.2 Varner, J. E. "Seed development and germination," in *Plant Biochemistry*, ed. J. Bonner and J. E. Varner. New York: Academic Press, 1965.
9.3 Toole, E. H., et al. "Physiology of seed germination," *Ann. Rev. Plant Phys.*, 7 (1956), 299–324.
9.4 Koller, D., et al. "Seed germination," *Ann. Rev. Plant Phys.*, 13 (1962), 437–457.
9.5 Wareing, P. F. "The germination of seeds," in *Vistas in Botany*, III, ed. W. B. Turrill. Oxford: Pergamon Press, Ltd., 1963.
9.6 Ikuma, H., and K. V. Thimann. "The role of the seed coats in germination of photosensitive lettuce seeds," *Plant and Cell Physiol.*, 4 (1963), 167–185.
9.7 Chen, S. S. C., and K. V. Thimann. "Studies on the germination of light-inhibited seeds of *Phacelia tanacetifolia*," *Israel J. Bot.*, 13 (1964), 57–73.

Appendix: Methods of Specimen Preparation

Photomacrographs

With the exception of Figure 37 (taken with a Pentax 35 mm camera) and Figure 146 (taken with a 63 mm Luminar lens on Panatomic X sheet film), all of the photomacrographs were taken with micro Tessar lenses onto Ilford N.40 plates, using a simple plate cassette and bellows. The background was a piece of black velvet (Figure 38 was taken against wet black filter paper). Illumination was from two desk lamps, fitted with frosted bulbs. Kimwipes(T) were hung in front of the bulbs to further diffuse the light. For Figure 47 the specimen was immersed in boiled water and illuminated by a fine pencil of light from a microscope lamp, reflected back across the specimen from a mirror. When it was essential to prevent desiccation, the specimens were mounted on a black insect pin to a wad of plasticine, stuck to the inside of a glass staining dish. A little water inside the dish, covered with a sheet of clear glass, ensured 100 per cent humidity during preliminary setting up of the specimen. The glass sheet was removed just prior to exposing the plate. For instruction in and help with these techniques we are indebted to Mrs. D. J. Carr, Botany Department, The Queen's University, Belfast.

Photomicrographs

Tissue Preparation. Figures 99–102 are photomicrographs of stained, freehand sections of fresh material. These sections were cut with Gillette Blue Blades.(T) The sections were first cut into tap water and washed for 1 minute. They were then transferred to an aqueous solution of 0.05% toluidine blue O and stained for ten seconds to one minute. Sections were then washed in tap water for one minute and mounted in tap water. The staining reactions between toluidine blue and the components of fresh tissue are similar to those described below for fixed material.

Figures 3, 24, 26, 33, 34 and 79 are photographs of living cells. Figure 40 is an autoradiogram of a section embedded in paraffin, and Figures 71, 72, 117, and 118 are whole mounts of cleared tissues. All of the other light photomicrographs are of sections from tissues fixed in acrolein, embedded in the slightly hydrophilic plastic, glycol methacrylate, and sectioned at 1–2μ with a glass knife on an ultramicrotome. The sections were dried onto a drop of sterile distilled water on a slide and stained as listed in the legends. The complete procedure, as well as details of the staining procedures, is given in Feder and O'Brien (1968). A short summary of the staining reactions follows.

Acid fuchsin and *fast green* are both anionic dyes that bind to positively charged groups in the cytoplasm. These dyes are strongly absorbed by mitochondria, plastids, and nucleoli. When used at low pH (in 1% acetic acid) and at high dilution (1:20,000) these dyes give excellent selective staining.

Along with *iodine*, they are useful in enhancing the phase contrast of various structures in these thin sections (see for example, Figures 4, 19–23, 29, and 30).

Toluidine blue O is a cationic dye that binds to negatively charged groups in the tissues. Although an aqueous solution of this dye is distinctly blue, it is a metachromatic dye that binds with carboxylated polysaccharides (for example, pectic acids) to give a pinkish purple color and to macromolecules with free phosphate groups (for example, nucleic acids) to give purplish or greenish blue colors. By a completely different set of reactions, toluidine blue stains polyphenolic compounds (for example, lignin and tannins) green, greenish blue, or bright blue colors. Hydroxylated polysaccharides such as cellulose and starch are not stained by this dye. Most primary cell walls (for example, those of parenchyma and epidermal cells) stain pinkish purple because of their polyuronide content. In lignified secondary walls (for example, those of xylem elements and phloem fibers) and primary walls impregnated with phenols (for example, those of some of the parenchyma cells of pea stem), the staining reaction of the dye with the polyphenols predominates. For reasons not clearly understood, the lignified walls of phloem fibers always stain bright blue whereas those of xylem elements stain a greenish color. The walls of cambial cells and some phloem elements are almost unstained by toluidine blue; the walls of other sieve tubes and companion cells stain pinkish purple. For further information on toluidine blue staining reactions see O'Brien, Feder, and McCully (1964) and Feder and Wolf (1965).

The *periodic acid/Schiff's (PAS)* reaction reveals the distribution of carbohydrates with vicinal hydroxyl groups (for example, starch, hemicelluloses, and pectins). Such groups are oxidized to aldehydes by the periodic acid, and these aldehydes form a strongly colored product with Schiff's reagent. For further information on the PAS test and for a discussion of the *Feulgen reaction* see Pearse (1960), Jensen (1962), Barka and Anderson (1965), and Lillie (1965).

Photomicrography. All of the photomicrographs of free-hand sections and plastic-embedded sections were taken with Carl Zeiss microscopes using planapochromatic lenses. Photographs in black and white were recorded on sheet film (either 2.5 × 3.5 inches or 4 × 5 inches) and a variety of emulsions (Plus X Pan. Contrast Process Pan, Panatomic X, and FP3) and appropriate developers were used. Photomicrographs in color were recorded on Ektachrome type B sheet film (using appropriate color-correction filters) and developed by Ektachrome process E3. Figure 3 was taken with a Carl Zeiss microscope equipped with Nomarski optics and a flash unit, and was recorded on Panatomic X sheet film.

Electron Micrographs

Tissue Preparation. Specimens prepared by ourselves were fixed in glutaraldehyde, postfixed in osmium tetroxide, dehydrated, embedded in Araldite, sectioned, and stained as described in detail in O'Brien (1967). The techniques used to prepare the specimens contributed by other authors are generally available from their published work.

Electron Microscopy. Our electron micrographs were taken at 80 Kv either with a RCA EMU 3F or with a Siemens Elmiskop I.

APPENDIX REFERENCES

Barka, T., and P. J. Anderson. *Histochemistry: Theory, Practice and Bibliography*. New York: Harper & Row, Publishers, 1965.

Feder, N., and T. P. O'Brien. "Plant microtechnique: some principles and new methods," *Amer. J. Bot.*, 55 (1968), 123–142.

Feder, N., and M. K. Wolf. "Studies on nucleic acid metachromasy. 11. Metachromatic and orthochromatic staining by toluidine blue of nucleic acids in tissue sections," *J. Cell Biol.* 27, (1965), 327–336.

Jensen, W. A. *Botanical Histochemistry*. San Francisco: 1962. W. H. Freeman and Co.

LILLIE, R. D. *Histopathologic Technique and Practical Histochemistry.* New York: McGraw-Hill Book Co., 1965.

O'BRIEN, T. P. "Observations on the fine structure of the oat coleoptile. 1. The epidermal cells of the extreme apex," *Protoplasma,* 63 (1967), 385–416.

O'BRIEN, T. P., N. FEDER, AND M. E. MCCULLY. "Polychromatic staining of plant cell walls by toluidine blue O," *Protoplasma,* 59 (1964), 367–373.

PEARSE, A. G. E. *Histochemistry: Theoretical and Applied,* 2d ed. London: J. and A. Churchill Ltd., 1960.

Index